SALMON

SALMON

A Scientific Memoir

JUDE ISABELLA

Rocky Mountain Books
www.rmbooks.com

Library and Archives Canada Cataloguing in Publication

Isabella, Jude, author
Salmon : a scientific memoir / Jude Isabella.

Issued in print and electronic formats.
ISBN 978-1-77160-045-3 (pbk.).—ISBN 978-1-77160-046-0 (html).—
ISBN 978-1-77160-047-7 (pdf)

1. Pacific salmon. 2. Pacific salmon—Symbolic aspects—British Columbia. 3. Native peoples—Fishing—British Columbia. 4. Indians of North America—Antiquities. 5. Pacific salmon—Ecology—British Columbia. 6. Human ecology—British Columbia. I. Title.

QL638.S2173 2014 597.5'6 C2014-904027-X
C2014-904028-8

Printed in Canada

Rocky Mountain Books acknowledges the financial support for its publishing program from the Government of Canada through the Canada Book Fund (CBF) and the Canada Council for the Arts, and from the province of British Columbia through the British Columbia Arts Council and the Book Publishing Tax Credit.

Canadian Heritage Patrimoine canadien

Canada Council for the Arts Conseil des Arts du Canada

The interior pages of this book have been produced on 100% post-consumer recycled paper, processed chlorine free and printed with vegetable-based dyes.

For my three muses: Tobin, Vaughn and Leo.

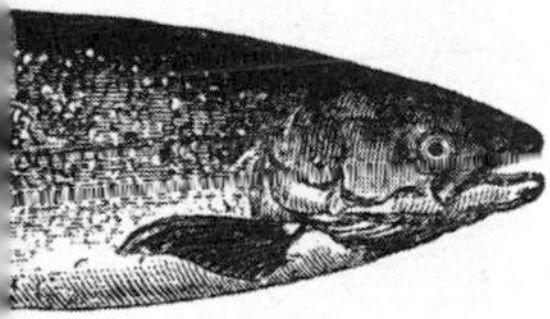

One of the reasons we gave ourselves for this trip – and when we used this reason, we called the trip an expedition – was to observe the distribution of invertebrates, to see and to record their kinds and numbers, how they lived together, what they ate, and how they reproduced. That plan was simple, straight-forward, and only part of the truth. But we did tell the truth to ourselves. We were curious. Our curiosity was not limited, but was as wide and horizonless as that of Darwin or Agassiz or Linnaeus or Pliny. We wanted to see everything our eyes would accommodate, to think what we could, and, out of seeing and thinking, to build some kind of structure in modeled imitation of the observed reality. We knew that what we would see and record and construct would be warped, as all knowledge patterns are warped, first, by the collective pressure and stream of our time and race, second by the thrust of our individual personalities. But knowing this, we might not fall into too many holes – we might maintain some balance between our warp and the separate thing, the external reality. The oneness of these two might take its

*contribution from both. For example, the Mexican sierra has "*XVII-15-IV*" spines in the dorsal fin. These can easily be counted. But if the sierra strikes hard on the line so that our hands are burned, if the fish sounds and nearly escapes and finally comes in over the rail, his colors pulsing and his tail beating the air, a whole new relational externality has come into being – an entity which is more than the sum of the fish plus the fisherman. The only way to count the spines of the sierra unaffected by this second relational reality is to sit in a laboratory, open an evil-smelling jar, remove a stiff colorless fish from a formalin solution, count the spines, and write the truth "*D. XVII-15-IV.*" There you have recorded reality which cannot be assailed – probably the least important reality concerning the fish or yourself.*

—JOHN STEINBECK,
The Log from the Sea of Cortez, 1–2

CONTENTS

Preface xi

Acknowledgements xv

Points of Interest xvii

Chapter One 1
The Salmon Doctors

Chapter Two 29
Noble Savages or Savvy Managers?

Chapter Three 65
Everything Eats Everything Else: Salmon in the Rainforest

Chapter Four 101
The Biological Black Box

Chapter Five 139
Life, Anthropology and Everything

Notes 183

Bibliography 193

PREFACE

My purpose in writing this story was to investigate a narrative that is important to the identity of the Pacific Northwest Coast of North America – a narrative that revolves around wild salmon. In the standard narrative, salmon are as central to our origins, desires and troubles as is the apple to the story of Adam and Eve. Salmon mean survival, health, pleasure and longevity. Yet this mythic narrative has always seemed too simple to me. We may dutifully exhibit our reverence for the fish that fed the First Nations of the Pacific Northwest, but how can the narrative guide us if all we really know is that salmon flood grocery stores in the fall and taste good? How can we – as consumers, as citizens, as neighbors of the ocean – more fully come to know the fish that supposedly defines the natural world of this place?

I conducted my research as a science writer, inspired by John Steinbeck's *The Log from the Sea of Cortez.* The best way to achieve reality, Steinbeck writes, is by combining narrative with scientific data.[1] So I went looking for something different from the story of constant conflict that most people read about in popular media. I searched for a narrative of the scientists who study the fish, either directly or indirectly, a story of the slow accrual of data and the sudden breakthroughs of insight that define our mounting, shifting, imperfect knowledge of salmon. Steinbeck was not writing about salmon in *The Log from the Sea of Cortez*, but his meditation on a fish called the Mexican sierra goes straight to my purpose here.

He describes the way we would come to know the fish in a laboratory, a lifeless, colourless, stiff, evil-smelling thing with a number for a name. It would be easy to count the spines on such a fish, much easier than on a boat during an expedition, especially if the sierra "strikes hard on the line so that our hands are burned, if the fish sounds and nearly escapes and finally comes in over the rail, his colors pulsing and his tail beating the air." In such a case, Steinbeck argues, a whole new "relational externality" comes into being – "an entity which is more than the sum of the fish plus the fisherman."[2] The "unassailable" reality of the lab is gone; counting the spines of a thrashing fish is undeniably difficult. But Steinbeck argues that the live encounter between fish and scientist or fisherman engenders a more important reality than the lab. Perhaps that is going too far; for most scientists, the lab and the field are happily entwined. But as a science writer, I tried to follow Steinbeck's example and trace the narrative journeys – the encounters, planned and unplanned – of scientists and citizens as they struggle to understand the world around us. Such narratives rarely make it into scientific journals.

I went on about a dozen field trips with biology, ecology and archaeology lab teams from the University of British Columbia and Simon Fraser University in Vancouver. I travelled with the federal Department of Fisheries and Oceans on board the Canadian Coast Guard ship the *W.E. Ricker*. I joined an archaeological crew from the Laich-Kwil-Tach Treaty Society on a dig in Campbell River, British Columbia. And while I journeyed with scientists, I was also reading up on fields as diverse as taxonomy, sustainability and folk biology. In particular, a 1938 PhD dissertation by anthropologist Homer Barnett from the University of Oregon, "The Nature and Function of the Potlatch," was riveting. A concise 2011 treatise on sustainable Indigenous systems by economist Ronald

Trosper at the University of Arizona, "Resilience, Reciprocity and Ecological Economics," expanded my view of what "economy" means.[3] And works by psychologist Douglas Medin at Northwestern University and anthropologist Scott Atran at the University of Michigan opened my eyes to the similarity between people who immerse themselves within an ecosystem for their entire lives and those who study it their entire lives.

My conclusions, after about four years of on-again, off-again research, beginning in 2009, surprised me.

ACKNOWLEDGEMENTS

My deep gratitude and thanks go to the First Nations who welcomed me to their territories: the Heiltsuk, Stó:lō, Sts'ailes, Tla'amin, and the Kwiakah, Xwemalhkwu and Wei Wai Kum (the Laich-Kwil-Tach Treaty Society). I want to thank Tyrone McNeil of the Stó:lō Tribal Council and his family for welcoming me to the family's dry rack fishery – though not in the book, their hard work and knowledge set the tone of the story. There were also the scientists and knowledge holders who welcomed me into the field, lab and/or answered all manner of questions: Will Atlas, Megan Caldwell, Tim Clark, Dee Cullon, Brooke Davis, Randy Dingwall, Erika Eliason, Amy Groesbeck, Scott Hinch, Yeongha Jung, Dana Lepofsky, Quentin Mackie, Iain McKechnie, Duncan McLaren, Rhy McMillan, R.G. Matson, Heather Pratt, John Reynolds, Christine Roberts, Anne Salomon, Carol Schmitt, Mary Thiess, Brian Thom, Marc Trudel, Michelle Washington, Louie Wilson and Elroy White. Eric Peterson, Christina Munck and the Hakai Beach Institute hosted me and unknowingly provided me with access to some of the final threads of the story. I would like to thank David Beers at *The Tyee*, who said "yes" far more often than "no"; the Society of Environmental Journalists for travel funding; the University of Victoria's anthropology department for providing funding, an academic home and proudly claiming me as one of their own; and the captain and crew of the *CCGS W.E. Ricker*. April Nowell and David Leach provided support, encouragement, comments and good-natured pressure to get the manuscript "done already." I especially want to thank the ever-patient Tobin Stokes for answering frantic phone calls from a lost traveller and giving really good directions, and for packing a house and moving almost single-handedly, sometimes with only a bicycle as transportation. Finally, I want to thank my editor Margaret Knox, whose voice I now hear in my head every time I write (thank you, Meg).

POINTS OF INTEREST

My field trips took me to points along the British Columbia coast: the Fraser and Harrison rivers (see diamond on the map), Tla'amin territory (triangle), Quadra Island, Phillips Arm, Campbell River (square) and the Central Coast (circle.) I also travelled on board a Canadian Coast Guard ship that travelled almost the breadth of the province's coastal waters.

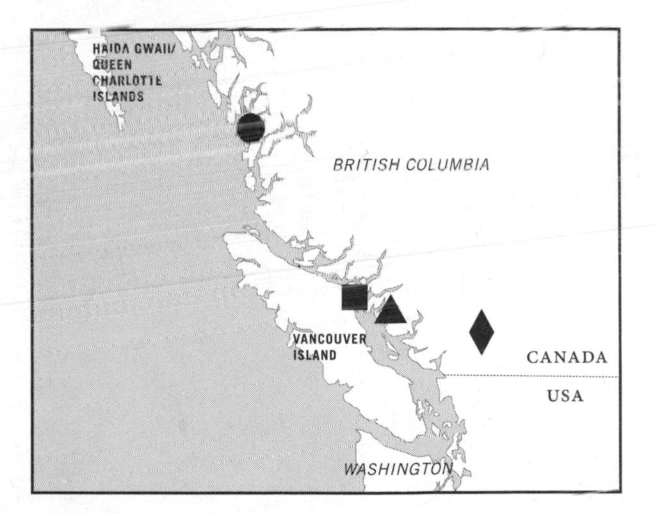

CHAPTER ONE

The Salmon Doctors

It is usually found that only the little stuffy men object to what is called "popularization," by which they mean writing with a clarity understandable to one not familiar with the tricks and codes of the cult. We have not known a single great scientist who could not discourse freely and interestingly with a child. Can it be that the haters of clarity have nothing to say, have observed nothing, have no clear picture of even their own fields?

—JOHN STEINBECK,
The Log from the Sea of Cortez, 1–2

IT'S DRIZZLY, COLD AND MUDDY, and a folding table on the south bank of the Harrison River in British Columbia is no place to perform open-heart surgery. Yet Tim Clark is attempting it. His tongue sticks out in concentration as he leans over his delicate patient, who is sucking in anesthetics through a tube down its throat. The patient's flesh is slippery, but he slices deftly into the chest cavity. Within minutes, he has stitched up the wound and handed off the patient to be carried away, slightly groggy but still kicking. Next please.

The van full of medical supplies behind Clark – gauze, forceps, gloves – is a MASH unit without a war. To the sockeye salmon resting on the operating table – a Rubbermaid container – the process must seem more like an alien abduction than surgery.

Clark is no alien, though he is Australian. His purpose is to insert a data logger into the cavity behind the gills of each fish, near the heart. Once the fish is released back into the Harrison River, a tiny computer will continuously record heart rate and temperature as the fish makes its final sprint to Weaver Creek, the natal stream where this population of sockeye will spawn before dying. Each surgery takes one to 15 minutes, depending on the fish's sex; male salmon have thicker ventral tissue and need fewer stitches to close the opening.

Clark and the rest of the scientists arrived early, in the cold mist of a September sunrise. Swaddled in fleece, rain gear and chest waders, they set up tents, tables, scalpels and tubes and waited for the fish. Then the fishers, men from the Sts'ailes First Nation's fishery program, started running the beach seine to catch patients for Clark and his colleagues.

It's high drama to watch them pull it off. They fix one end of the net to a truck on shore, the other to a motorboat that zooms across the river and loops back to shore, snaring the flopping, over-half-a-metre-long creatures. By 4 p.m., they'll have deployed the seine net eight times, catching fewer salmon as the rain stops, the sun shines, the day warms and the fish sink deeper into cooler water.

The scientists have partnered with the Sts'ailes fishers for the past six years, the fishers taking DNA samples for their own fisheries program, the scientists inserting monitors to track how fish physiologically manage to take so much punishment in their final push to keep their gene pools going. In most ways, the operation feels a lot like a traditional fish camp – except that these salmon are netted for data, not food.

Knowing nothing of seining, I jump in with everyone else to help pull in the netfuls of salmon. Being on the small side, I am the weak link in the tug-of-war. The fish slap my legs, tossing, jerking, and snagging their teeth in the netting. Standing in the midst of their thrashing gives me a sense of how powerful salmon need to be to swim against the river's current. It turns out hauling in seine nets is best left to hardier souls! I go to help Clark and a couple of graduate students. They scoop the freshly caught salmon in hand-held nets and make the transfer to the operating room's waiting area, the shallow waters at river's edge where they plop the fish into a pen. For a while, I stand thigh-deep in water and write down tag numbers as the scientists evaluate each fish, plucking off a scale to send to a Department of Fisheries and Oceans (DFO) temporary lab at nearby Weaver Creek.

Sockeye populations can be identified by scale patterns viewed under a microscope. Within an hour of sending the first catch, the lab, set up just for this purpose, calls Clark to tell him that 11 out of the 25 are Weavers; the rest are Harrison River fish. This distinction matters. Clark's study compares fish physiology among different populations within the sockeye species. The focus of the study is Weaver sockeye, not the more plentiful Harrison River fish.

Humans have known, through observation in the ancient past and through experimental science today, that the more salmon runs there are, the healthier the species is overall. Whatever challenges salmon face – climate change, disease, industrial pollution, overfishing, hatchery production, fish farms – they will ultimately evolve or go extinct, depending on their diversity. Yet scientists are forced to prove over and over again, in deepening detail, that a species is doomed without population diversity. This is especially true as the climate changes and waters warm. The work that Clark and his colleagues do shows the fine, unseen differences among sockeye populations. It should be simple. But it isn't simple because our relationship with sockeye is overwhelmingly about money. And there is nothing simple about money.

The 150-metre stretch of land along the Harrison River where Clark and the others conduct fieldwork belongs to the Sts'ailes First Nation and is called simply "The Park." Roughly five kilometres from the Fraser River, it's one of the most productive fish habitats in the Fraser Valley. All five Pacific salmon species – pink, chum, chinook, coho and sockeye – swim these waters, traditionally running from June to March. Even today, after years of commercial fishing, logging and industrial pollution, the ecosystem erupts with life. The fish attract loads of birds. In the next couple of weeks, the Park will swarm with teals and other ducks. During the past six or seven years, cormorants have made a big splash here too, the Sts'ailes say, gobbling any fish that fit into their bills, including one-kilogram trout. An occasional sea lion has been glimpsed trolling the Park, having travelled 15 kilometres from the Pacific Ocean.

Clark's riverside heart-surgery project is one of many in-depth sockeye studies in Canada. Fish biologists Scott Hinch and Tony Farrell at the University of British Columbia in Vancouver and Steve Cooke at Carleton University in Ottawa manage most of them. Lift the lid on their research and it's like picking up a patio stone and seeing a colony of ants at work, all frantically moving toward individual goals that converge on a single purpose: to understand the physiology of salmon in excruciating detail. No function goes unnoticed, it seems, from heart rates and temperature tolerance to aging.

A few strides away from Clark's station, one of the younger team members, Samantha (Sam) Wilson, stands under a tent and eviscerates dead sockeye, plucking out their brains and hearts. She flash-freezes the organs in liquid nitrogen and stores them in a cooler to be couriered overnight to Ontario. Wilson, an undergraduate student at Carleton University, is intrigued by a question of colour. She wants to know if brighter-coloured salmon age more slowly than dull-coloured salmon. A salmon's bright-coloured skin comes from carotenoids (antioxidants) in the food they eat. It's possible that brighter–coloured salmon (with higher antioxidant capacity) are better at preventing aging. It's possible that they survive longer on spawning grounds. If so, do they pass their antioxidant capacity to their offspring? The question arose from studies showing that birds with bright-coloured beaks tend to have higher antioxidant capacity and stronger immune functions than birds with dull-coloured beaks.[4] And the brighter the male beak, the more likely he is to interest a female, the implication being she will have healthier offspring if she goes for flash.

Wilson expertly cuts into a fish brought to the operating table by Graham Raby, who is charged with giving Wilson fresh kill. Raby, a graduate student at Carleton, evaluates Fraser boxes and fish bags. Fraser boxes are used by commercial fishers to revive fish nabbed as bycatch, nontargeted fish caught in the seine net. The boxes are painted black to soothe the accidental catch, and fresh water runs through them. Fish bags are essentially black duffle bags with mesh at each end, a low-tech method that conscientious sport fishers use to accomplish the same end – something they're not required to use yet. But Raby doesn't use these tools to revive and release accidental catch. Instead he attempts to revive his sockeye – and makes notes on whether he succeeds in reviving them – then bonks them on the head and brings them to Wilson for her study. The sockeyes are slippery and strong enough to launch a heavy lid off a Fraser box, so Raby weights the boxes with large rocks. Holding down a sockeye and administering a deadly blow on the first swing is tough, but Raby is quick and efficient.

At this point, after six years of study, this group's focus has been on stress physiology and the effects of temperature, particularly the effects of warming waters. The conclusion, so far, is what one might expect: fish that experience high temperatures naturally are better at coping with stressors under high temperatures than fish that experience cooler temperatures naturally. It's a bit like comparing Vancouverites on dry, 30°C days to visiting Torontonians. The Vancouverites are wilting, while the Torontonians are delighted to have escaped the humid heat back home. Although it's not exactly the same. We warm-blooded humans can handle it even if we don't like it. Cold-blooded salmon can't.

At 4:00 p.m., the buzz of activity is muted at the Park. The fishers are packing up their gear and the catch at the lab tents is dwindling. A warm breeze carries a sweet, hay-like smell from the grassy riverbank to overlay the odour of blood wafting from Wilson's fish morgue. She has placed 28 salmon brains in vials today for transport and later study. I imagine a FedEx delivery to the wrong doorstep, someone expecting smoked wild sockeye fillets, not raw fish brains.

At the surgery table, Clark continues at a feverish pace. "It's a girl," he calls, incising a belly with quick strokes. He inserts a data logger, encased in the same silicone used in biomedical implants for humans. Clark often adapts the tools of medical doctors for his fish studies. In his lab he has a meter originally intended to measure hemoglobin levels in human blood. He has recalibrated it for fish blood. Likewise, to count fish glucose levels at the lab, he uses a glucose monitor developed for diabetics.

The implant he has just inserted in the female sockeye will rest against her organs, and the tiny computer inside it will record internal temperature as well as electrical pulses from the heart. Once patched together, she'll go back to the temporary pen in the river for a few hours before she's let loose to find her natal stream. Not a single fish of Clark's has died since I arrived at dawn. They will live to spawn and then die. Naturally.

In a few weeks, Clark will go to Weaver Creek to find his tagged fish and remove the computers. The data should tell him if the Weaver fish look for the cool spots in rivers – called thermal refuges – to save energy. It should also show him how the fish allocate energy during migration. "No one really knows that," Clark says.

Erika Eliason does similar fish physiology work to

Clark's, but she doesn't contend with seine nets and river currents. A few weeks before tagging along with the crew working on the Harrison River, I drove down to see Eliason at the DFO lab at Cultus Lake, about an hour's drive east of Vancouver. All of the researchers at Cultus Lake, the same scientists at the Park, put in long summer days on fish studies at the lab, and though the setting is tame, the work is in many ways as strenuous as hauling nets on the Harrison River. The lab includes freshwater pools that dot a fenced, concrete area. At one pool, "chasers" take turns using their arms to churn the waters. The object of their flailing is a single adult sockeye, caught from the lower Fraser River a few days earlier.

To "chase" fish, you need a stopwatch, knee pads and lots of energy, particularly if it's a hot day. The chasers – four women, in this case – whip up a froth in the pool as another watches, a stopwatch in one hand and a clipboard in the other. "Way to go!" she calls, like a coach at a kid's soccer game. "Keep it up, keep it up!" After three minutes of being "chased," Eliason holds the fish up in the air to simulate what happens when a fish is caught. Then she places the stressed fish in a pool next to the one from which she plucked it, where its stress levels are monitored. The water temperatures in the pools range from an ideal 16°C to a worrisome 21°C. This one is at 18°C. The group will do this for 12 fish. As in Clark's fieldwork, a few questions are in play: How well do stressed fish recover? Does temperature matter to recovery? Does handling the fish change the aging process, making them less likely to reach their spawning grounds?

Everyone breaks for lunch and I follow them to a bungalow, a home they share for most of the summer as they conduct their fish studies. Eliason sits on a couch, seat cushions caving in with age, holding a coffee mug. She is cruising toward the conclusion of her PhD and is here to help fellow students with whatever needs doing, whether chasing fish, recording data or teaching fish surgery. I have 20 minutes to interview her, which is perfect; it's like sitting through a private TED Talk. Eliason could make an 8-year-old care as much about Fraser River fish distinctions as they do about superheroes.

If things get too warm, Eliason explains, some sockeye populations are likely to die of heart failure during their heroic journey to reproduce. "It's such a narrow part of their lives," she says. The migration lasts two to four weeks, but it's critical to the survival of the species. In general, migrating sockeye suffer when temperatures are above 18°C, she says, but not all sockeye populations will suffer equally. Each population, though the same species, is adapted to a specific river ecosystem.

The Fraser River sockeye populations are as diverse as the First Nations Peoples – and the early settlers – who could wrest a fortune, or not, from the mighty river. Whether they toughed out the climate and stayed or failed and moved on, the people left their names with the rivers the fish still swim: Stuart, Nechako, Quesnel, Chilko, Adams, Weaver and Gates.

The Chilko – historically about a quarter of the Fraser River sockeye run – are the elite athletes, physiological freaks with big hearts, incredible oxygen uptake and an ability to swim powerfully in waters as warm as 22°C. They start to lose steam at warmer temperatures, but they're still moving at 26°C. At the mouth of the Fraser River, the Chilko swim through warm summer waters, but by the time they've made their way up the river, they're cruising comfortably through chilly currents to their home 650 kilometres upstream, the glacial-fed Chilko River. They can handle the cold and the heat.

Nechako salmon migrate farther than Chilko – more than 800 kilometres up the Fraser River – but in water without temperature extremes. Heat the water up to 20°C and they stop swimming. They have the aerobic scope but not the heart of the Chilko. And Weaver sockeye, which travel a mere 100 kilometres or so to spawn, are the skinny weaklings on the beach. They have neither big hearts nor the Olympic athlete oxygen uptake. Compared with the mighty Chilko, they would have a tough time adapting to a warmer world.

It took Eliason three years to figure out these differences. She spent many evenings inserting catheters into sockeye and taking blood samples as they swam in fresh water pumped into big swimming tunnels made from PVC piping at the Cultus Lake lab. The goal was to compare how well they took up oxygen from the water to feed their muscles, both at rest and while swimming; this is what she calls "aerobic scope." She was also measuring how well they pumped blood around their bodies and their heart size. As she monitored the fish, she tweaked the water temperatures.

"Every stock is different," Eliason says, waving her coffee mug in the air. "The same rules aren't going to apply." Eliason keeps saying the same thing in different words. It's as though she's in a race to understand the breaking point of each sockeye type before the fish are destroyed. More importantly, she's in a race to make the world understand it and change its practices accordingly. "How is that [understanding] going to be put into practice when they're all in the river at the same time and you can't see who is who until you look at the DNA?" I ask. She shrugs and shakes her head. Best not to think about it too hard.

Unfortunately for salmon, especially the sockeye in the Fraser River watershed, habitat is more than a scientific concern. It's a commodity. The Fraser River is home to more than 100 sockeye populations with a commercial worth of more than $1-billion annually, on average.[5] Canada's commercial relationship with the fish is, of course, older than its scientific relationship. Since the Hudson's Bay Company began exporting salted salmon in cedar barrels from Fort Langley on the Fraser River in the 1830s, the number of people invested in sockeye has only climbed, while sockeye numbers have declined.[6]

To keep evolutionary pace and avoid extinction, a sockeye population needs to be big enough to encompass plenty of differences among individuals. To avoid catastrophe – a mystery disease, climate change or both – it helps to view each stock as if it contains the seeds of future diverse populations. It's comforting to know that it can take only 55 years for a salmon population to become reproductively isolated. In other words, split a population into two and within 13 generations they're on diverging genetic roads, widening their genetic heritage.[7] That's about the same number of years as only about two human generations. Less comforting is knowing that this happens only under the right conditions, when a population is adapting to a new, salmon-friendly environment, not to a rapidly changing, salmon-hostile environment. Not to the warming climate, in other words, of the greenhouse-gas era.

Sockeye salmon generally swim up the Fraser River in their fourth year of life. They lay eggs and die, and in spring the fry emerge. Most fry head for a lake, probably to avoid predation. When a fry emerges, it's only about the size of an inchworm, a perfect snack for a bigger fish. To a young salmon, the ocean would hold the same attraction as a buffet for a growing human adolescent, but sockeye fry must opt for leaner rations. In food-poor lakes, they have less to eat, but they're also less likely to end up as lunch. A trade-off – food availability versus predation – and no doubt a good evolutionary move.

To find the lake, fry rely on either the sun's position or polarized light patterns. Put fry in a covered round tank to deny them visual cues, odours and water currents, then rotate the magnetic field with a direct electrical current, and the fish will navigate by Earth's magnetic field.[8] The sensory-deprived fish head in the direction they normally would to their home lake. Chilko River fry, for example, which enter Chilko Lake from the north end, will orient south.

The good news for sockeye fry in the lakes is that they're usually the most abundant fish feeding on tiny crustaceans. The bad news is that just as for the adults that come back to spawn, temperature matters. Raise the water temperature and the fry's metabolism kicks into high gear so that food barely maintains them. Lower the temperature and it takes the fry longer to digest. It makes sense, then, that when food is plentiful, the fry grow best at 15°C, and when it is less plentiful, they grow best at lower temperatures.

Of course, sockeye nursery lakes vary in

temperature, elevation and geography, and individual populations vary in their adaptive responses, just as humans do in varied habitats. For example, Andeans, Tibetans and Ethiopians living at altitudes above 2500 metres have three different biological adaptations to oxygen-thin air. Andeans have more hemoglobin, the oxygen deliverer, in their blood. They can breathe at the same rate as a person living at sea level, yet move more oxygen around their bodies. Compared with people who live at sea level, Tibetans take more breaths per minute. Scientists think they also use another naturally occurring gas, nitric oxide, more efficiently to widen blood vessels and allow more blood to flow. The Ethiopian adaptation is known to be different but remains a mystery.[9]

All of the human populations that have adapted to higher elevations likely did it culturally at first, using fire and warm clothing. They were buying time, culturally, to allow biology to catch up. Other animals lack that evolutionary luxury, and salmon have the added complication of spending their lives in more than one habitat. They move from fresh water to salt water to fresh water again and have wildly different needs at different life stages. Food, for example, is not a need at all once they start their final migration to spawn, but this means they have a lot of eating to do before starting up-river. Over 98 per cent of the weight most salmon gain over the course of their lives happens in the ocean.[10]

Clark, Eliason and everyone working at the Park and Cultus Lake are amassing data to add to our collective knowledge about salmon. Yet on a wider scale, their studies are about us. Studying the aging process in fish, for example, expands our knowledge of the human aging process. Antioxidants and bright colouration? The findings probably won't translate directly, but along with lots of other small scientific breakthroughs, the studies of aging in fish populations will contribute to a better understanding of human health. On a practical level, the work of scientists at the Park already helps fisheries managers, at the very least, understand how complicated it is to manage this kind of resource. Maybe, eventually, the data will affect federal fisheries policies. But once they release their data, scientists know they have little, if any, control.

"We all know, from the cod collapse on the East Coast, that even some of the best science can be ignored," Scott Hinch told me as we stand outside the students' summer home at the Cultus Lake lab one afternoon. The northwest Atlantic Ocean cod collapse in 1992 devastated Canada's Maritime provinces. Fishers who grew up thinking they would carry on family traditions, woven through multiple generations, lost livelihoods forever when the cod population plummeted from 500 years of overfishing.[11]

Hinch doesn't think it's some vast conspiracy. "I don't think there's any pattern," he says. "I think it really depends on the local situation and the people who are involved. History has shown us that in other fisheries small studies can provide really unique information, that if the right people see it and understand it, they can act on it quickly."

Generally speaking, the East Coast lobster fishery, depending on the jurisdiction, is a good example of science informing policy. Lobster fishers are restricted in the number of traps they can set and the kind of traps they can use, and they set free lobsters that are too small or too big. At one time, the common practice was to keep the biggest lobster, but then scientists showed that big lobsters were mega-breeders. It was an easy change to make and it helped restore lobster populations.[12]

A University of Toronto graduate, Hinch

studied warm water fish in Ontario lakes before capitulating, over 20 years ago, to the faunal charisma of salmon – they're big, heroic and pretty swimmers that taste good – and a chance to live in British Columbia. Hinch might not use the word "capitulate," but as a biologist he knows that faunal charisma often comes with cash. Even when an animal is extinct, faunal charisma can keep funds flowing. What else explains the drive to clone a woolly mammoth?

What Hinch worries about most when it comes to salmon are two horsemen of the environmental apocalypse: warming temperatures and pathogens. The Fraser River is close to 2°C warmer than it was less than ten years ago.[13] For cold-blooded salmon, that's a problem. "Warmer temperatures are going to have a big influence on disease proliferation," Hinch says. Yet the research to understand it hasn't been done.

All sorts of circumstances drive pathogens – bacteria, fungi and infectious agents such as viruses and prions (a cause of the fatal brain disease bovine spongiform encephalopathy or mad cow disease) – to morph or spread. Human practices are implicated in more ways than one. Crowded fish farms in Chile, for example, hastened the spread of the infectious salmon anemia virus in 2007 and the country continues to deal with the disease.[14] And climate change, as always, is a big player in pathogen behaviour. So given the almost slam-dunk certainty that Earth will keep warming in our lifetime, what can we expect for sockeye?

A study by DFO scientist Kristina Miller – co-authored by Hinch, Farrell, Cooke and others – provides a worrisome hint of things to come. Miller's study published in 2011 exposed a possible disease killing Fraser River sockeye before they had a chance to spawn. Referred to as "salmon leukemia," it is potentially the culprit behind two decades of falling salmon numbers, culminating with a 2009 collapse, when only a million fish came back to the Fraser River out of an expected ten million.[15]

Today, the immune systems of many sockeye salmon are compromised, possibly by a virus. Scientists don't know how it's transmitted, whether from parents to offspring or fish to fish. They don't know whether it's endemic to all fish or only to salmon. They're not sure if it's related to climate change or to other stressors. What they do know is that if a fish has the signature of the possible virus, it is most vulnerable to sickness when morphing physiologically to make the switch from salt water to fresh water. Late-run sockeye that migrate in early autumn when temperatures are cooler – and that have the signature of the possible virus – exhibit a fatal behaviour change, a timing issue. Since about 1996, they've been migrating anywhere from three to eight weeks earlier than normal.[16] Late-run salmon that show up early to spawning grounds are more likely than others to die before they can reproduce. Hinch and Miller are supervising research – heating sockeye and pink salmon and observing their cellular response – that could soon yield answers about the relationship of the possible virus to warming waters.

Rising temperatures have already been blamed for the *Ichthyophonus* parasite that, since the 1980s, has been infecting and killing Yukon River chinook salmon (the river is almost 6°C warmer than it was less than ten years ago, in the early 2000s). "Ich" (appropriately pronounced "ick") was thought, at first, to be a fungus, and the state of Alaska dragged its feet in addressing the problem. It took independent research to convince Alaskan fisheries officials that in warm years, Ich was infecting almost half the Yukon River female chinooks. After the worst years of 2003 through to 2004, the infection has steadily declined, to about

4 per cent. But Yukon River chinook numbers have kept plummeting – by 60 per cent from 268,537 in 2003 to 2011 when only 107,000 were left.[17]

Not all chinook are equally susceptible to Ich: lab tests have shown that some BC chinook populations might be less susceptible than those of the Yukon.[18] In Washington state's Puget Sound, the chinook have low levels of Ich, even though they feed on heavily infected herring. They also resist infection from a Yukon River parasite if introduced. The resilience of certain chinook populations may lie in their diversity. University of Washington scientists in 2010 analyzed over 50 years' worth of research on sockeye salmon from Bristol Bay, Alaska – the largest sockeye fishery – and showed that genetic diversity gave the fishery its stability. They called it the "portfolio effect": just as a diverse financial portfolio ensures financial stability as markets go up and down, a diverse genetic portfolio gives a biological system stability.[19]

Like most places fished today in British Columbia, the Park is within a First Nations traditional use area. Aboriginal Peoples have fished here for thousands of years. They're known, in recent centuries, as the Coast Salish, a group bound by language and culture. Their communities stretch from the northern Lower Mainland to Vancouver Island and Washington state. The Sts'ailes, the fishers who run the beach seine for the studies, are Coast Salish. Archaeologists have found Coast Salish settlements on both banks of the Harrison River, and on mid-river islands, all built within five metres of the river or of sloughs. Settlement dates remain unclear, but it's safe to say human occupation of the Park is ancient.

Humans have lived along the rugged, fjord-riddled coast of British Columbia for at least 11,000 years.[20] They've corralled fish to their doom with nettle fibre nets, stone traps and wooden weirs. The evidence is there, from California to Alaska, even on rivers and streams that no longer host salmon runs. (A few kilometres upstream from the Park, at Morris Creek, the remnants of a wooden weir were still visible in the 1970s.) Archaeologists believe that the size of the runs of salmon was less important to early peoples than was access to many populations, big and small. People needed protein in every season; they fished probably through most of the year but through a strict set of rules. They also depended on hunting mammals and managing flora from the ocean to the mountains. It's possible that the clues to sustainable management lie in this not so distant past.

Unearthing answers will require cooperation among scientific disciplines ranging from biology to anthropology. That's a stretch: biologists and anthropologists tend to have starkly different mindsets. For biologists, inferring a conclusion – for example, venturing what effect fish farms might have on wild salmon – is dicey. In some cases it's okay, but the mantra remains: correlation is not causation. Anthropologists and archaeologists, on the other hand, are more comfortable with inferences, even when others use their data to come to wild conclusions about humanity. A number of studies focusing on societal collapse, for example, have been treated as conclusive evidence that as a species we are incapable of conservation.[21]

There's feedback at play in much human-focused science. Confronted with a fact, such as the disappearance of big mammals at the beginning of the Holocene epoch in North America, the spotlight on the human role can be relentless. Egos entangle with theories, and powerful proponents boost their own speculations. Charles Darwin's breakthrough – evolution by natural selection – gave biology the grammar to move forward as a science. It also gave cultural anthropologists a headache for the past 150 years. Both biology and culture help to explain human behaviour, and it remains a challenge to consider them both complementary and separate.

We're all the same, and we're all different – and that's the starting point. As a species, we all have the same biological needs. It's how we meet those needs that differs. For most of the 20th century, biologists, ecologists and archaeologists found plenty of evidence of past environmental destruction. They didn't find much evidence of environmental conservation.[22] The problem was, until recently, we didn't really know what conservation in the distant past might look like.

CHAPTER TWO

Noble Savages or Savvy Managers?

It is not enough to say that we cannot know or judge because all the information is not in. The process of gathering knowledge does not lead to knowing. A child's world spreads only a little beyond his understanding while that of a great scientist thrusts outward immeasurably. An answer is invariably the parent of a great family of new questions. So we draw worlds and fit them like tracings against the world about us, and crumple them when we find they do not fit and draw new ones.

—JOHN STEINBECK,
The Log from the Sea of Cortez, 137

IF DEE CULLON HAD LIVED in the 19th century – and if she'd been a man – she might have lit out for the territories as a coureur de bois or ship's navigator, braving wind, ice and mud to gather furs, compass points and stories to tell. Cullon, a slender, delicate-boned, 37-year-old anthropologist, gets to spend much of her time these days slogging around forbidding terrain on the west coast of Canada. Her tools, though, are a good deal more sophisticated than the sextant or theodolite of the early explorers.

"I use Google Earth a lot," Cullon says. "I zoom in and look at the estuaries because, for whatever reason, the photos are taken in summer ... and during the days with very low tides." For the past six years, Cullon has been searching for fish traps, ancient ones that will tell her something about the people who made them and something about their environment. It takes her years to sift from her data the clues she seeks to social, legal and ecological puzzles. Yet gathering the data is a simple matter of slogging through wet and cold – much like a coureur de bois.

"Do you know what you're looking for?" she asks me on a raw August morning in 2010 at Menzies Bay, an old fishing and logging camp midway up Vancouver Island's east coast, just north of Campbell River. It's my first day out with her team. She asks it kindly, head cocked and eyebrows raised, as though welcoming a bewildered kindergartner to the first day of school.

"No idea."

We are looking, it turns out, for the round tops of wooden stakes, darker than the sand and roughly the diameter of a coffee mug. As waters move, they bury all kinds of evidence. But they also erode sand and mud, exposing evidence. Some years, Cullon's team finds signs of elaborate fishing ventures almost every place they visit: jumbled stones and stakes, lines of stones and stakes, stone traps alone, wooden stakes alone. I assume the head-down stance of the archaeologist, eyes on the ground, walking, walking, walking.

Christine Roberts, another cheerfully long-suffering outdoorswoman, shorter than Cullon, hails us to her square of beach, where she has found one, then two more, stakes, all in a row. The wind whips Roberts's dark hair as she digs around the stakes, measuring diameters and distances and planting a tiny red surveyor's flag. The forlorn little flag reminds me, for some reason, of Henry VII: "... and I will see that the English flag is planted in this distant land."[23]

It is not, of course, for Henry VII that Cullon and Roberts labour but rather for the Laich-Kwil-Tach Treaty Society who want to learn all they can about fish weirs and traps in traditional territories of the Kwiakah, Xwemalhkwu and Wei Wai Kum First Nations. What kind of wood are the stakes? How many are there? What are their shapes? How do they overlap with historical data? What kind of fish did they trap? How old are they? To whom did they belong?

For the tribes, it's partly a matter of law. The weirs are "fences" in the intertidal zone, below the high-tide line. A fence traditionally signifies ownership in British common law. So a weir "fence" is a potential trump card for coastal First Nations wrestling the government over territorial claims. Roberts, as well as Randy Dingwall and Louie Wilson, two other crew members who are scouring the beach further down, are Laich-Kwil-Tach members.

"The Canadian government makes a big deal about property and fences," Roberts says, as she turns a stake over in her hands while Cullon snaps rapid photos. "These are our fences, they prove we had technology below the high-tide line."

Later I ask the men about the fish traps. Dingwall, the silent type, simply says with a smile, "They're great." Wilson, the youngest of the three, says the traps are so important to asserting their ownership of the foreshore but they also give the people a sense of their own history. "Growing up as a kid, I didn't know about the traps at all, and..." he trails off, turns his palms up, shakes his head, a puzzled expression crossing his brow, then continues, "I don't know why."

If nothing else, the traps prove the tribes' deep ties to the land.

Cullon does a GPS reading and makes notes on the stakes. The heavy hand of industrialization has hit this area hard and she is glad to find any signs of ancient fishing. The beach is packed flat and looks scraped. Menzies Bay was a log yard for much of the last century, with camps, a railroad and steam-run equipment sprawling along the shores and into the forest. One of the largest-known, human-made explosions on the planet tore off the top of Ripple Rock, just around the point in Seymour Narrows, in 1958, because, lurking close beneath the surface, the rock had ripped into more than a hundred hapless boats.[24] A couple of wood pilings still poke out of the water from a pier dating to Menzies Bay's busy industrial era, but all that's left of the factories is a single log-sorting and wood-chipping facility near town.

Legal questions aside, ancient fish traps and weirs are cryptic guides to much older ways of relating to the coast. For generations of Native people – for thousands of years – the fisheries were a complex, integrated solution to meal planning. The people of the First Nations, sophisticated scholars of the coastline, dedicated themselves to harnessing protein from the ocean to feed as many people as possible, for as long as possible, with as little effort as possible. Today's menu planning, though, is more complicated. It's a Rube Goldberg contraption of interconnecting stovepipes, siphoning fish to competing interests: industrial fishers, commercial fishers, sport fishers, subsistence fishers – separate parts never forming a whole.

As part of her work, Cullon has been asking Indigenous Elders where they used to fish, hunt and pick berries, where the old villages were, and the old burial sites. She has been creating, more or less, a narrative map of the land and seascapes.

That information, compiled in a traditional-use study for the treaty society and cross-checked with marine charts and Google Earth, helps Cullon and a colleague, archaeologist Heather Pratt, pinpoint where to throw their efforts into the search for ancient fish traps.

Part of what draws archeologists to the field, long before they specialize, is the notion that all humans need the same things and that, across the globe, humans often go about acquiring stuff like food in similar ways. Fish traps turn up practically everywhere you find anadromous fish – such as salmon and herring – that migrate from the sea to spawn in rivers or close to shore. This is also true for their opposites, catadromous fish – mostly eels – that swim from lakes and rivers to spawn at sea. The Thames River's medieval fish weirs, the Maori people's eel weirs in New Zealand, the Amazon's ancient, 500-square-kilometre weir in Baures, the Passaic River's pre-Colombian weirs in heavily urbanized New Jersey, various European weirs dating to 8,000 years ago – all these display similar ingenuity for the gathering of similar dinners.[25]

The Pacific Coast of North America is no

different. How people used the weirs depended on the fish, the materials available, the broader environment and the human culture. But from California to Alaska, fish traps lined the ocean shores for thousands of years and their remains are easy to find – provided you know what to look for. One may be stone, one wood, one a wall, one a fence. Just as the people of Amazonia – in that sprawling, tropical, fisheries complex – intensively domesticated the landscape, so too did the people of the Pacific Northwest Coast – including British Columbia. In the nearby Tla'amin territory, for example, archaeologists have found what they think are stone fish traps interspersed with clam gardens. In the same way that gardeners of domesticated plants accede to the whims of their mini-ecosystems – moving blueberry bushes, peonies or whatever is growing poorly until they find a spot where the plant thrives – so must ancient peoples have moved their fish traps until they found the sweet spots. The more experience you have with the land or water, the more abundant your garden or harvest of seafood.

In North America, weirs have given us many place names. "Toronto" is likely a variant of the Mohawk word *tkaronto* that means "where there are trees standing in water." A fish fence. That place was about 125 kilometres north of the city, where Lake Simcoe empties into Lake Couchiching. Near the town of Orillia, Ontario, today, this place is known as Mnjikaning in Ojibway and means "at the fence."[26] Mnjikaning was where thousands of specimens of the now endangered American eel and various fishes were trapped and harvested continuously for almost 5,000 years. The Mnjikaning fish weirs are still there.

The abundant productive capacity of the old weirs is referenced in such sneering colonial documents as the 1827–1830 journals kept at Fort Langley, the Hudson's Bay Company outpost in British Columbia. Shortly after the fort was built on the Fraser River, the company's chief factor wrote that the "lazy Indians" couldn't be bothered to hunt beaver for pelts. The darn salmon was so plentiful that the Aboriginals hardly needed to work. The colonials may not have understood that the bounty of the weirs was a work in progress – a work that had begun thousands of years earlier. But after derogating the Natives, they apparently learned to recognize a good thing. In the 1830s the work of Fort Langley morphed quickly from trading fur to trading salmon.[27]

The day after my inaugural stake-walk at Menzies Bay, we putter across, in a hired boat, to Phillips Arm, a remote estuary some 200 kilometres northwest of Vancouver. Cullon's aim, on this enormous mud flat, is to find wooden stakes she had spotted during an earlier survey. From the rocky beach, we can't see the scars of the heavy-duty logging of the past century; any remnant tree stumps are hidden behind a second or third growth of forest. Nor do we know, at a glance, that gold was mined nearby at one time. In the milky light of early morning, the estuary looks pristine. Yet both of these get-rich-quick industries altered fish habitats here forever.

The mud slurps and sucks at our legs as we walk onto the estuary, grabbing as high as the tops of our knees until we reach firmer ground at the middle of this intertidal zone. It doesn't take long to find the tops of wooden stakes, laced in seaweed, poking up from the tidal flat. No trowelling is necessary; the estuary is not packed flat like the scarred Menzies Bay. Pratt crouches to pick seaweed off a stake. A light-brown ponytail pokes through the back of her baseball cap; a red bandana protects her neck from the wind and sun. Beyond her, gulls float in a tidal pool. They're uncharacteristically silent – almost regal – without French fries or other human leavings over which to fight.

It's chilly, and the mist mutes the voices of the crew as they work at removing four of the dozens of stakes. The team looks like Lilliputians extracting teeth from a giant, salivating mouth. It's hard work, and they grunt and struggle, ladling up muck that is sucked infuriatingly back down the hole almost as quickly as it comes out. Roberts, Wilson and crew member Rhy MacMillan have known each other for a while, as students at the Vancouver Island University anthropology program. They pull together with the stolid equanimity of draft horses – until the first stake erupts from the mud, and they whoop like children on a fun ride. They've got at least one prize, one specimen to haul back to the lab for dating, and for identifying the tree species from which it was carved.

Another three stakes come out along with the sun, and I wander over to watch the crew's surveyor Ken MacPhail. I've been wondering about the plunger-shaped protrusion from his backpack. McPhail's glasses give him a professorial air, but his sandy hair and fair skin – weathered from years of outdoor research – cancel the effect, and the plunger cocked beside his head might best set him in a science fiction movie.

MacPhail's contraption turns out to be the antenna of a roving GPS Total Station. He is there to electronically record and map the exact position of the stakes. If they disappear from view again, the GPS recording will direct future researchers back to the site. McPhail will feed the data into a computer program to generate a map of the weir. This nontactile method saves hours for archaeologists, who used to work with tape measures, compasses and prisms. Whether high- or low tech, maps can reveal patterns, and the patterns in the maps of these archaeologists reveal human engineering in the intertidal zone.

As at Menzies Bay, early Indigenous people at Phillips Arm developed weirs to lead fish from their watery habitat to human plates, and their fishing continued uninterrupted for thousands of years. The weirs coaxed, rather than ripped, food from the shore. Estuaries like Phillips Arm, where ocean tides tumble into river mouths, ease smolts – salmon kids – into sea life. Some salmon species and populations are ready at an early age for the salty ocean. For them, the estuary is but a quick stop on the way to a proper meal. Other salmon species stay for months to bulk up; a bigger juvenile of such a species has a better chance to survive in the open ocean. Still others use the estuary to adjust to the saltiness of the sea in a less abrupt version of a newborn baby's struggle to gasp air into fluid-filled lungs. They linger in the mixed waters, moving up and down a few metres, adjusting.

As recently as a couple hundred years ago, Phillips Arm, like many estuaries, was a gentler home to smolts. Forests perched on the coastline regularly sloughed off mature wood and debris that splashed into the water, creating a structurally complex habitat in which small fish could forage and escape predators. The waters were turbid, creating a protective smoke screen that hid vulnerable smolts from such sharp-eyed foes as trout, sculpins, blue herons and older salmon. The kids played their lethal game of tag in the equivalent of a little park playground with a jungle gym, rather than in an open field in which the biggest and fastest would dominate.

I leave the techies recording data on the estuary and turn to the mud-grubbers and their freshly harvested stakes. We each sling one over a shoulder to take back to the lab for dating and sampling. They're each about a metre long and waterlogged, thicker than a walking stick and heavier than baseball bats. The wood might be hemlock, cedar, fir or pine. Cullon thinks the fish traps from Phillips Arm probably date to about 2,000 years ago, as do those from three other nearby sites. These include the Nanaimo estuary, which Cullon has found to be dense with traps – solid wood posts made of hemlock stakes with cedar slats strung between them. That time period – about the time of the birth of Jesus in Nazareth – is key in West Coast archaeology, because, as in the Middle East, lots of things were changing. Some archaeologists think this was when populations grew, houses got bigger and people stored more food. Why exactly this was happening is tough to answer, but the society encountered by Europeans a few hundred years ago was a very old one, dating back at least 2,000 years.

We head toward the river mouth, where the silhouette of the wiry Randy Dingwall is striding back and forth rapidly, looking for more stakes. A former tennis champion, Dingwall has the lean shape and moves of an athlete, though he is a 50-year-old smoker. "The man in the distance," Louie Wilson calls him; Dingwall is beside you one minute, and when you turn your head, he's gone, as though he could teleport 500 metres.

Through the fractured lens of tidal relics, Cullon's team is scrutinizing long-ago meal-gathering practices, but their research bears on the future of the coastal fisheries. Traps and their fish remains tell archaeologists something about resilience, about the way in which ancient communities protected themselves from the perniciousness of Mother Nature. Trap technology was ubiquitous for so long – people modified a template so well to suit a place – that these coastal cultures testify to the wise, intimate management of a local resource.

Traps illustrate the human capacity for thinking in the long term; they allow for selective fishing and for the passage of fish to spawning grounds. In the old days, the Phillips Arm salmon got to reproduce in large enough quantities to sustain their population, and people got to eat for generations. As diverse habitats and diverse fish species changed, the traps adjusted with them. Today we tend to use a technology until there are no fish left. Think cod and factory ships.

Too many people went after cod with too efficient a technology, driven by a market economy. But it's too easy to say that. It also lets us off the hook. If too many people were eating cod, it's because too many people occupy the planet, and no one can wave a magic wand and make billions of people disappear. That's the kind of overgeneralization that paralyzes us. So it pays to look more closely at one particular problem, like how few chinook return to Phillips River each year. Can that be changed?

All five species of salmon come back to spawn and die up the Phillips River. In a few weeks time, in September, about 300,000 pinks, the most abundant and smallest of Pacific salmon, will swarm the Phillips estuary. Yet only a few hundred chinook (called spring or king in the United States – Canada officially changed the name in 1965) will come. If you're going to call yourself the Salmon Capital of the World, Campbell River's claim to fame, that's not good enough.

Since the 1980s, the Gilliard Pass Fisheries Association, a non-profit salmon enhancement organization, has taken about 15 per cent of the Phillips River chinook run, aiming to harvest 200,000 eggs to rear in hatcheries. Once the fish in the hatcheries reach smolt stage, they're released, either into the Phillips estuary or about six kilometres upstream in Phillips Lake. But like most hatchery-raised chinook in British Columbia, the survival of hatchery-raised Phillips Arm chinook is crushingly low, less than 1 per cent.[28] The explanation for such poor numbers flows back to genetics.

Chinook populations, like sockeye populations, are tremendously diverse. Contemplating smolt size is about as meaningful as looking at a ballroom full of women from around the world and coming up with some kind of average. Some populations produce a statuesque, robust smolt, a Venus Williams, while others produce a slender, gracile smolt, a Lady Gaga. A grown female fish has anywhere from 2,000 to 17,000 eggs to deposit; Phillips chinook females tend to produce a respectable 5,000 eggs. More important is the question of how long they stay in fresh water, if at all, once hatched. Ocean-type chinook leave fresh water for the salty sea immediately, though they will linger in the brackish estuary. Stream-type chinook stay in fresh water for as long as three years. During their residence, the stream types are up for a fight and flaunt colourful fins; their growth seems to be tied to the light cycle. They grow more slowly than ocean types, initially, but they are bigger at the moment when they enter the ocean. Their destinations are different too. Stream types head for the central north Pacific Ocean and return in spring and summer, while ocean types migrate along the coast and return later, mostly in summer and fall. Cross the two, and the stream-type genes play second fiddle to the ocean-type genes.

The point is that each has adapted to a different ecological niche. Above the 55th parallel, chinook salmon are almost all stream types. Below it, scientists think that ocean types dominate. Efforts to enhance chinook habitat have relied on raising ocean-type fish. But does it come down to something as simple as a boundary, a line in the sand?

We walk the Phillips estuary toward the sea, the long-limbed Dingwall, as always, far ahead. When we're closer to the mouth of the river, we turn inland, toward a side channel. (The estuary is quite huge and it's hard to tell exactly how close we are to the main stem, but there are lots of side channels pouring in.) From ahead of me, someone lets out a whoop – a stake juts out of a dry stream bed, a lone sentinel waiting for our team to relieve it of its centuries-long duty. But no, it's not alone: more stakes cling to the riverbank. Did they support a platform for spearing fish as they headed upstream to spawn? Wilson and Macmillan head further upstream, heads down, scouting. Pratt, Cullon and

MacPhail catch up with the rest of us, lugging their high-tech map-makers. But we should leave soon; the tide is coming in fast. Cullon notes the number of stakes – 23 – and takes a quick GPS reading. We call back Wilson and Macmillan. It's more than time to leave.

Running is out of the question. Clumps of seagrass, probably an accidental import, trip up even Dingwall as we hurry, our feet tilting side to side, never finding a smooth surface, a surface that will soon to be the ocean bottom. By the time we find more solid footing, the water is gushing into the estuary, knee-high. Dingwall, MacMillan and Wilson bound across the onrushing tide. Dingwall surges ahead, glancing back briefly and waving his arm, urging everyone to go faster. The rest of us are rushing to keep up. I find myself balancing on my left foot, water at thigh level, desperate to keep from plunging the cameras on my back into the tidal flow. I stretch to reach my right foot and pull a sandal back on over a thick, sopping wet, wool sock. By the time I straighten up and regain my balance, the three fleet men have disappeared around the point. I hurry and catch up to Roberts. A few more steps and she appears to hit a hole, the water rising to her chest. I hold my backpack above my head and start inching toward her. I pause and we look at each other. Roberts shakes her head and points behind us to the bank we're skirting. "I think we'd better head up." We turn back, along with the other three who haven't made it. One by one, we scramble for high ground, passing off our stakes and heaving each other out of the water. I had no idea archaeology could be so thrilling.

"This is grizzly country you know," Roberts says for the third time that day as she hauls herself up a steep incline thick with trees and brush. The ripe wild blueberries and huckleberries are gone by early August, but the thought doesn't give me much comfort; grizzlies also munch on green leafy plants and roots. We stumble through the brush, all gripping waterlogged stakes, which have no use as climbing poles, and the full packs on our backs catch in the undergrowth. I'd like to think it was the expensive, cumbersome gear that slowed our group in the rushing tide of the estuary. But even without our loads, we probably couldn't have kept up with the others. I imagine those three lucky souls lolling in the sun as we struggle back, taking the long way around, forging an untrodden route. At the lip of a ridge, Roberts hands off her stake to the person above her and grabs at a shrub to pull herself up. I pass her the stake I'm holding and do the same. We look, in vain, for a path.

"I have never had this happen before," Cullon says, bemused. She radios the boat operator to try a new pick-up point. Not possible: the skipper is afraid of running aground. We take turns leading the way, pushing aside brush, careful not to let it whip the face of the next in line, tripping on understory thick with salal, climbing over tree trunks to stay as close to the water as possible, passing our stakes endlessly back and forth. The bluff is high and steep, messy with undergrowth and logs; one slip and I'd tumble to the water, if not impale myself on a tree branch. After almost two hours, we figure out a place where the boat can pick us up and lurch within view. A huge cedar log stretches across the forest floor between us and a possible path down a lower stretch of cliff. I'm in the lead and the only thing to do is climb over the log. I stretch myself over it, arms too short to place the stake on the other side. I drop the stake, figuring to pick it up once I've hoisted myself over the log. The stake starts to roll. And roll. And roll. I scramble over the log in time to see it bound down the bluff and break into two pieces. One by one, the group slides over the log. They bunch up behind me, silent.

Oops.

Cullon picks up the pieces of what may be a 2,000-year-old relic. "You know what," she says, "it happens." She smiles. "These traps are everywhere."

We're exhausted by the time we meet the guys – and the boat – at the point. We take turns boosting each other to the deck, Cullon last. She boosts Pratt. The boat shifts, and Cullon lands in the mud and water.

"Sorry!" Pratt shouts, safely on deck. Cullon laughs and splashes aboard for a cold ride home.

I ask Dingwall how, in the rush across the river mouth, he managed to outpace the two younger men, Wilson and MacMillan. "I took off my boots as soon as I realized how fast the water was coming," he says. "I was barefoot."

Experience, in other words, won out.

On day three of my fieldwork inauguration with Cullon and her crew, we're at Deepwater Bay on Quadra Island, tramping down a creek, feeling skunked, seeing only steel cables from logging years. They run the length of the stream, like phone lines to nowhere. Mid-morning, after walking for more than three hours, we find ourselves back at the beach, dutifully walking transects, though we've already criss-crossed the sand many times. It seems odd not to find stakes. Deepwater Bay is known to host plentiful numbers of sockeye on their way to the Fraser River. In the 1920s, canneries operated a commercial trap at Seymour Narrows, where the fish waited for the tide to change before heading upriver.

"I found one!" Roberts calls, finally. We're slow to react, waiting for a punchline. Wilson finally gets to work, guessing where the stakes might line up and looking for the next one, wiggling his gumboots in the sand. Wilson's "boot" technique is infallible; he finds two more stakes quickly, and someone starts singing, "May the circle, be unbroken…" We end up flagging a big trap – 43 stakes in a curving line – most of them thanks to Wilson's gumboots. The little red flags flutter in the breeze.

"I like seeing all those flags," Cullon says.

Pratt laughs. "That's because you're so controlling."

"Hey," Cullon says, "if I weren't so controlling, we wouldn't be out here." Cullon is meticulous about organizing her team, recording their data and planning. Her report will be as exhaustive and labour-intensive as a doctoral thesis.

Later as we perch on a jumble of big grey rocks,

waiting for the boat, Roberts and Cullon discuss British royalty, particularly King Henry VIII and his obsession with sons. But before long, they're back to musing about fish traps. I mention some stone weirs I read about in a master's thesis about the traps of the Heiltsuk First Nation on British Columbia's Central Coast, a few hundred kilometres north. One weir stretches 100 metres or more.[29] That's a lot of heavy rocks, a lot of wood to carve into stakes and slam into the ground.

"They would have had a lot of fish," Pratt says. "Everyone must have worked, the entire village." We stare silently at the ocean for a few moments. It rises and falls, as if breathing.

Fish weirs have probably been around since the Paleolithic era – the Stone Age – capturing the bounty of the ocean shore. The technology wasn't always used benignly or in balance, especially in medieval times. In the mid-1300s, fishers in the Alps paid an archbishop 27,000 white fish and 18 lake trout annually for the right to catch more and wiped out a population of white fish in a generation.[30]

In general, when a social order changes and economic interests compete, fish weirs become a problem for lawmakers. The Magna Carta of 1215 banned all inland traps in Britain.[31] Medieval noblemen in continental Europe routinely destroyed peasant traps. In the 18th and 19th centuries, industry silted, dammed and polluted rivers and streams across Europe, rendering freshwater weirs unproductive.[32] As seafaring technology advanced, fishers hauled their catch from the open ocean instead of the shores. In Britain, Parliament moved to protect the property of the landed gentry (including fish) in 1861 by banning all fish traps not in use before the Magna Carta.[33]

Upon arriving in the Pacific Northwest of North America, colonists assumed that the vast amount of fish being trapped illustrated the abundance of the region, but they also assumed that fish traps overexploited the resource. By that time, any memories of European fish traps were associated with waste. They used both perspectives to get what they wanted. The area was in an upswing of productivity in the early 1800s when European trappers and settlers arrived – it was an all-you-can-eat buffet for sea lions, sea otters, salmon, herring, zooplankton and humans. What the colonists didn't know – and Indigenous people did – is that the ecosystem, fed by what scientists now call the "chaotic" North Pacific, was notorious for its pendulum swings of abundance and scarcity. If an animal, humans included, was to survive in the North Pacific, it had to adapt to chaos.

The colonists, understanding neither the newly "discovered" societies nor the North Pacific habitats, failed to realize that Aboriginal trap management had sustained a healthy fishery in a chaotic environment for thousands of years. Or maybe they simply didn't care. The Canadian government banned First Nations fish traps in the late 19th century; there was no room for the older social order and economic system.[34]

Catching salmon is the heart of the economy in Campbell River, BC. To keep Campbell River alive, the Gilliard Pass Fisheries Association works hard to keep the salmon coming back. That's been no easy task in the past decade, and they've finally decided to try something new. Maybe that simple line in the sand doesn't exist when it comes to types of chinook. After decades of disappointing chinook runs, the fisheries association has enlisted help from the most vocal chinook salmon breeder on the coast: Carol Schmitt.

I first chatted with Carol Schmitt in the summer of 2011. The following spring, I drove to her private hatchery, Omega Hatcheries, on Central Lake on Vancouver Island. Just as Cullon might have been a coureur de bois if she were a man a couple of centuries ago, Schmitt might have been an erudite naturalist, writing elegant, witty letters about her fish experiments. Schmitt is in her mid-50s, her angular features softened and her skin weather-grooved, but like many natural scientists, she has the heart – and the curiosity – of a 10-year-old. Growing up near Central Lake, Schmitt used to rescue salmon fry stranded in the dry years when rains skirted the island and dried up the creeks.

Schmitt is convinced that the reason chinook enhancement fails is because hatcheries pump out ocean-type fish, accelerating growth by stuffing them with food, coddling them in water warmer than their natural environment and releasing them as smolts after only eight months. At that tender age, their immune systems can't fight off fatal ocean viruses. In the estuary, they're likely to die from a disease they're too small to weather, and even if they stay healthy, odds are they'll lose a game of tag with a predator.

Schmitt, a graduate of a two-year fisheries-technician program at British Columbia Institute of Technology, has worked with fish for over 30 years. Over 20 years ago, she began to change her chinook rearing conditions.

"You want to see the smolts," she says, shoving her chair back from the table where her computer sits in her mobile home, where she has been guiding me through reams of her data. She grabs a baseball cap, tucks her long blonde hair inside it, closes the buttons on a flannel jacket and leads me out the door. The walk to the hatchery pools, at Schmitt's brisk stride, takes a couple of minutes from her door.

The fish she's taking me to see are Phillips smolts, a stream-type chinook that the Gilliard Pass Fisheries Association has hired Schmitt to raise. The association and Schmitt both have high hopes for the experiment – the association because they'd like more chinook, Schmitt because she likes fish and would like her ideas to be vindicated.

In the pools, Schmitt tries to mimic the rugged conditions the smolts will have to survive when they are dumped into various rivers. The Phillips River smolts will go in May. They're a swarm circling a pool with colder water, less food and slower growing conditions than most hatcheries ever dare to provide.

"Call it adaptive management," Schmitt says

as we first swing by a large concrete storage shed behind her home, stopping to grab a bucket of feed and a scoop. Schmitt has been imposing her tough-love conditions on various types of chinook, and she's willing to work with anyone who wants to measure their survival rates once they are released into the wild and head back to spawn. "If something isn't working, I ask myself, what can I change? I can't control the ocean, I can't control the climate. I can only control how I raise the fish."

If the stream-type and ocean-type juvenile life histories are not genetically based, but rather reflect environmental conditions experienced during early juvenile rearing in fresh water, Schmitt has a good point. And the association and federal government hatcheries have released the wrong fish for decades.

Schmitt cannot resist showing me everything on the way to the pool of Phillips smolts: salmon, salmon, more salmon and novel fish others pass on to her for observation and experimentation – albino coho, blonde rainbow trout and sturgeon. We stroll inside another building with round pools. The albino coho could be a curiosity in a roadside attraction – anyone who finds, or accidentally breeds, something odd tends to bring it to Carol.

"I think I qualify as an official fish geek," she says. "Hello, fish," she says, voice rising as she leans over a tank with such motherly warmth that I half expect one of the fingerlings to jump up and wag its tail like a puppy.

We head outside and meander among a dozen or so round, white covered pools, a slice of each roof open so Schmitt can feed the smolts. When she approaches the pool holding the Phillips brood, the fish swarm to the opening, and I think again of eager puppies – thousands of teeny pets sensing someone with treats in her pocket. "Here you go," Schmitt calls, scooping the feed and flinging it over the calm surface of the pond. The surface explodes in a noisy froth of silver bodies. They arch out of the water, curving this way and that, flicking their slender tails, thwacking the water into a tiny maelstrom. Schmitt was careful to toss the feed as evenly as possibly, giving every smolt a chance. But they're not about to take it easy and risk seeing the nearest mouthful, a dry pellet of mostly fish meal, fish oil and soy, snatched away by a sibling.

While I watch, the image of a baby pool packed with guppies in a suburban backyard suddenly tugs at my memory. A scene from a movie about piranhas then flashes across my mind. But as the smolts' wriggling intensifies, it is wild herring, another silvery little fish, that swipes away any other analogy. I once witnessed a school of herrings spawn, their slender bodies flashing in the sun as they swarmed the shallow waters of the West Coast like a giant superorganism, laying eggs, squirting sperm, roiling the waters for hours as I watched. Thousands of smolts feeding are a lot like herrings when they spawn.

In a few weeks, Schmitt will take these voracious baby chinook, raised to be tough enough to handle an unpredictable ocean environment, via truck to the Phillips River. It will take a few years to find out if she's right, if hatcheries should be raising more stream-type fish. Like most long-term ecological research, it requires time and patience and a willingness to be wrong. Gilliard Pass Fisheries Association will keep track of the data. If Schmitt is right, hatchery-raised chinook may have a fighting chance if hatcheries across the Pacific Northwest practice the kind of adaptive management Schmitt advocates. It might convince an industry that a cheaper investment – it's quicker and less costly to raise ocean-type chinook – is a deadly mistake when it comes to wildlife. More research showing that mimicking the wild, not controlling the environment, supports an old idea that to improve on Mother Nature, humans need to pay attention to what other species are telling us. The implications of Schmitt's research on this one fish are profound.

It doesn't necessarily take a PhD to break ecological ground. Schmitt, with only a diploma from a Vancouver tech school, has something that worked just as well for such luminaries as Mary Anning, the English woman who collected fossils with her dad and eventually gave the world evidence of dinosaurs, and Gregor Mendel, a monk who played with pea plants and gave us genetics. Like them, Schmitt has an open mind. After decades of raising fish, she brings a playful, what-if approach to the job. She starts with a specific, observed fact: chinook smolts rarely live to see adulthood. From there, she generalizes. Something happens to them when they get to the ocean, where they spend most of their time. Generalities are not true all the time, so simply reasoning this way is flawed. It could be that chinook, no matter what type, do not do well as hatchery fish at all. So scientists often turn to deductive reasoning. Science mixes inductive and deductive reasoning, though a mix of the two methods is not purely the domain of scientists.

In this case, Schmitt started with a limited number of simple statements, such as hatchery chinook fail to thrive in British Columbia, and when smolts are held longer and are grown bigger, they seem more robust. What would happen if these more robust salmon were used to enhance a fishery? By teaming with the association, Schmitt might get a chance to prove, deductively, that she's on the right track. Since it's just one study, she would have to replicate this a number of times to offer proof to other hatcheries. Now if the government decided to get on board with her theory, proof would come much sooner as more hatcheries would try Schmitt's method. With a return rate of 1 per cent for hatchery-raised chinook, what's to lose?

About a year and a half after our excursions in quest of fish traps, Cullon sends me her report. The stakes we slung over our shoulders for the circuitous trudge out of Phillips Arm were hacked, it seems, from balsam trees over 1,000 years ago. A case could be made, then, that the fish traps and weirs on the coast were always legal – that the First Nations had fences all over the coast, proof that they managed and cultivated the land in a way a colonial government would have to recognize. Although they went steps further, into the intertidal zone and beyond, their management expanded to the fish circling the Pacific Ocean. They decided which fish lived, which fish fed them, the relationship a reciprocal one that kept both species happy.

The Laich-Kwil-Tach tribes can claim a precedent predating the Magna Carta. Traps built and in use before the 13th century were grandfathered in when King John signed the document. Canada's West Coast wasn't claimed by Britain until the end of the 18th century, but those "fences" were in use well before the Magna Carta, the document described as "the greatest constitutional document of all times – the foundation of the freedom of the individual against the arbitrary authority of the despot."[35] By enclosing their fisheries with fences, Indigenous people claimed ownership of an ecosystem. No one had the right to take that away.

So, as Roberts and Dingwall and Wilson would argue, the Crown should have honoured their ownership. Everyone knows going back to the old ways is impossible. But a claim of ownership in a place where Indigenous people waded through the legal landscape with their hands tied and eyes blindfolded for centuries signals an intellectual victory and a path to re-establishing a give-and-take relationship with the ecosystem. The original weirs might be impractical in the modern era but the idea of resilience is not.

It's no secret today that Indigenous people worked the land and seascapes. It's figuring out how they gathered food so successfully for so long that has archaeologists looking back through a different lens, looking for chances to redraw this ancient world to perhaps find a way to shape ancient knowledge into a modern, sustainable practice. This kind of archaeological work – racing the tide, drawing maps, pondering resources – would be impossible without the ecologists.

CHAPTER THREE

Everything Eats Everything Else: Salmon in the Rainforest

The true biologist deals with life, with teeming boisterous life, and learns something from it, learns that the first rule of life is living.

Everything ate everything else with a furious exuberance.

—JOHN STEINBECK,
The Log from the Sea of Cortez, 25, 41

A SILVER TIN CAN of a vessel comes chugging toward the dock, a boat that would look more at home in a bathtub than on the fickle waters of British Columbia's Central Coast. It skirts the wing of a float plane tied to the dock, and I can see, squeezed into the wheelhouse at the back, two young scientists wearing big sunglasses. I ask John Reynolds, an ecologist from Simon Fraser University (SFU), what the funny little boat is called.

"*The Tickler*," he answers, and I burst out laughing. He shrugs, laughs and grabs a rope as the boat bumps alongside the dock. "I have no idea. You know, kids." The kids – a group of graduate students and post-docs studying wild salmon ecology – use the awkward-looking but trusty tub in their far flung adventures. Each year they fan out across 50 watersheds on the Central Coast, 700 kilometres north of Vancouver, mostly within the territory of the Heiltsuk First Nation. They come in shifts, depending on their project, staying from spring until the salmon die in the fall, teasing apart the salmon food web, a web that was woven at the end of the last ice age. The two in the boat, graduate students Noel Swain and Jeanette Bruce, are part of the Reynolds Lab and Earth2Ocean Group at Simon Fraser University in Vancouver. Reynolds holds the Tom Buell BC Leadership Chair in Salmon Conservation and Management, funded by NGOs and private and corporate donors.

We've driven from the airport, about five minutes away, to meet the students at this sturdy, federal government dock at Bella Bella, the Heiltsuk First Nation reserve on Campbell Island. *The Tickler* has just enough room for Reynolds and me to jostle onto a cold, metal bench for a ten-minute ride to the lab's HQ, an old cannery on Denny Island. It's sunny but we put on survival jackets, a shield against the wind that ruffles Reynolds's black, curly hair; traces of grey clues to his age, early 50s, though he looks younger. Traces of an English accent, however, belie the fact he was born and raised in Toronto, the accent an artifact of the time he spent at the University of East Anglia in the United Kingdom.

The water we skip across to Denny Island is part of the Inside Passage, a marine highway for BC Ferries and other shipping traffic. Most of the boats are shuttling between Port Hardy, at the northern tip of Vancouver Island, and Prince Rupert, on the mainland and a few hundred kilometres north. To the travellers on them, the scene is a spectacular backdrop, a single frame in the long movie of their passage, a holiday snapshot, perhaps, on a cruise to Alaska.

To these lucky scientists, it is much more. They return again and again to this bull's eye at the middle of a wet ecoregion that runs from southeastern Alaska to northern California. It is a landscape in constant, if slow, motion. The Central Coast teetered and tottered in a tectonic rhythm throughout the last ice age, sinking and rising as glaciers compressed and retreated. It is a geographic linchpin for geologists and archaeologists, and as I return with the scientists over half a dozen times, starting with Reynolds in the fall of 2010 and ending with a visit in the fall of 2013, I'll come to think of it as a metaphoric linchpin in this particular salmon story.

Swain pulls *The Tickler* alongside a serviceable dock at Denny Island, not as elaborate as the government dock but well maintained. A blue building, the old cannery, looms above. We shoulder our packs and head up the gangplank to unload.

Reynolds's work is complex and simple at the

same time. From a purely theoretical standpoint, he's trying to get at a key concept in ecology: How do salmon affect the population and diets of their consumers? Are they a mega-dietary supplement for Pacific wrens, freshwater sculpins and for such shrubs as salmonberries, which hug the banks of the rivers and streams? Because the fish overwhelm a nutrient-limited system once a year, the evidence of their impact should be plain to see. And it is, but only when Reynolds gets the more complex part right. His challenge is to design experiments that can uncover exactly how – and how deeply – salmon are integrated into their coastal surroundings.

He scrutinizes population densities of animals and plants along salmon-spawning streams compared with nonsalmon-spawning streams. Then he analyzes them for salmon's signature dietary contribution – nitrogen levels – gleaned through isotopic analysis at the lab. Since 2006, the Reynolds Lab has been drawing an increasingly detailed image of coastal BC's salmon-fed ecosystem.

The sun is out and high enough, after we've unpacked, to allow a visit to one of Reynolds's streams on Denny Island about a 15-minute boat ride away: a place called Gullchucks. It's the start of spawning season for chum and pink salmon, the main fish that Reynolds studies. In another week, more students will join Swain and Bruce to walk streams and count fish. "If we see coho we'll count them," Reynolds says over the roar of *The Tickler's* engine and the wind. "But we see very few of those. They come later. Besides, they're very secretive, hard to find. We see the odd carcass now and then." Likewise, chinook and sockeye are rare; sockeye young need to spend a year in a large lake, and fish can't get to the lakes here, Reynolds says. "We see the odd stray occasionally but they're not really part of our story."

At Gullchucks, Swain drops us close to shore, then consults a tide chart and steers *The Tickler* away, anchoring in deeper water. He climbs into an orange kayak to paddle back and join us at the stream's mouth. Wearing chest waders and boots, we walk toward the forest, pausing on the beach to look at a stone arc maybe 100 metres long, a remnant of a trap, the first stone trap I've seen. It looks like a whale's mouth, flushed clean by thousands of tides. A raven swoops overhead, coming to rest in a Douglas fir tree. The raven is a "trickster" in traditional First Nations' stories of the coast, mischievous, and, like humans, equal parts foolish and smart. In some stories, the raven disturbs the force that keeps the ocean level, resulting in the tide that, twice a day, brings food flooding in to the "people," human and nonhuman.[36] That's a lot of power to attribute to the trickster – launching the cycles that bring phytoplankton washing into the intertidal zone, feeding clams and mussels, which in turn feed crabs – and bringing salmon to get stuck behind rock walls for easy pickings.

We haul ourselves a little further upstream in chest waders and boots, and I find three wooden stakes peeping out of the mud along the bank. "Has anyone rebuilt a fish trap to see exactly how they work?" Reynolds asks, bending down to study the stake.

"Not that I know of," I say. "It's illegal – technically." Reynolds's eyebrows shoot up. In 2013, a team of scientists and First Nations built a traditional wooden fish trap on the Koeye River about 40 kilometres south from where we stand.

Swain, his aviator glasses and short afro at odds with his waders, starts to look for salmon. He leads the way, sure-footed after two seasons of scrambling through the temperate rainforest for 12 hours at a time. Bruce, with a reddish tint to her hair and a sprinkling of freckles across her small nose, looking as if she should be dancing and singing at a Celtic festival, clumps through the sedge behind Reynolds, talking about plants.

"Sedges have edges, rushes are round, grasses are hollow right up from the ground," Reynolds chants. He hurries along the riverbank after Swain, and Bruce and I, on our first day in the field, lurch along after them, planting our feet carefully on the uneven ground to avoid slipping into cold water.

Swain and Bruce, in their 20s, are East Coast exports – New Brunswick and Newfoundland, respectively – and the disappearance of the cod fishery is part of their psyche. Many of Bruce's friends who would have followed their fathers into cod fishing left the Rock for work in the Alberta oil sands. Swain's father, a DFO scientist, learned the hard way what happens when politics trump science: he watched helplessly as poor government management strategies, based on politics and not science, failed to protect the cod fishery.

"What's this?" Reynolds asks Bruce, fingering a feathery plant that looks soft enough for a pillow. "Lanky moss," Bruce answers, without hesitation. He keeps asking about mosses until he trips her up. It's a game with them, Reynolds pointing out plants and trees, Bruce answering like a contestant on Name that Plant: Amabilis fir! Douglas fir! Salmonberry! Huckleberry! She's a champ, even in this strange habitat, a continent away from her craggy Atlantic-island home.

We scramble across the stream and loop back to the seashore, finding another stone trap. Its lower, crumbling walls suggest an older time frame than the first, the whale's-mouth arc. We pause, wondering again what it would take to rebuild a trap and catch salmon the old way.

At the river's mouth, the early arrivals to their natal stream crowd each other. Some males jockey for position next to a female, an occasional splash the only sound to interrupt the hum of flowing water. It flows so strongly that sometimes a male fish loses his place entirely and is sucked back toward the ocean. He fights again to move forward. We're mesmerized, but the shadows lengthen and it's time to head back.

We've counted one of those wily, secretive cohos – and 512 pinks.

Of Pacific salmon, pinks are the puniest, averaging just over two kilograms in weight. They have the shortest life cycle, too, a scant two years.[37] In the Fraser River, their migration peaks during odd years, but on the Central Coast – again, something of a linchpin – they return in abundance during even and odd years alike to more than 130 streams and rivers. Even-year pinks, it turns out, edge out odd-year pinks, but not by much. The short lifespan of pinks means they probably can adapt more quickly than other species to environmental changes. They're like colonizer species, alder trees or aspens that shoot up in a clear-cut before anything else. Yet they are the least-enjoyed salmon, the type people used to call "cat food."[38]

Pink salmon leave their freshwater homes right away, so they're less affected than other species by habitat degradation in streams and rivers. They can migrate through fast moving water when they have to. Scientists say pinks are better swimmers than sockeye, although fishers will argue that point, and they can stand heat better than other salmon species. Pinks might even prefer warm water. In general, larger, older salmon are found in colder waters and smaller, younger salmon in warmer waters. Japanese lab studies have concluded the bigger the fish, the increased preference for colder waters. But compared with sockeye of the same size, which wilted in warm waters in the Japanese studies, pinks are fine with a little heat.[39] Their tolerance for heat is a big deal, it turns out, because what comes with a moderately warm ocean is favourable conditions for the survival and growth of most sub-Arctic zooplankton species. In other words, a seafood buffet for pinks.

Russian studies that measure copepod (tiny crustaceans) biomass in the Bering Sea have found that warming ocean waters enhance ecosystem productivity from the lower trophic levels (toward the bottom of the food web). This is especially true for planktonic crustaceans, which is good news for pinks. Pinks eat a lot of pteropods, a planktonic sea snail. And records from Russia note that during cooler periods – the 1950s and 1960s – regional pink stocks in Russia were low. In 2009, the year that Fraser River sockeye failed spectacularly, pinks showed up en masse.[40]

Here's the other rub for other species: pinks spawn in the lower parts of rivers and within weeks they're off to the races. They enter the ocean early, get to the table first and stuff their faces. If a warmer ocean shifts production of food earlier, pinks will chow down even earlier than their cousins. Luckily for the other species, pinks are terrible at finding their natal streams. But wait, that's not so lucky for other species after all.

Because maturing pink salmon are more likely than other salmon species to get lost, they redistribute themselves and their genes liberally among spawning regions. Fish that show up in some other population's stream are called strays. They're crucial to strengthening, or even re-establishing, a salmon population. The most dramatic example of the influence of stray salmon showed up after Washington's Mount St. Helens erupted in May 1980. Salmon that had spawned in nearby Toutle River the previous fall died when mudslides and ash choked the river. Strays from those Toutle River populations were the only ones to leave a genetic legacy. It's a good reason to protect all salmon spawning streams – one population might end up repopulating more vulnerable watersheds. But it's also a reason why pinks might outcompete other species in a warming world.

By the way, freshly caught and barbecued pinks are a taste sensation. Probably thousands of times tastier than cat food.

When Reynolds, Bruce, Swain and I emerge from the bush, the stream banks glow orange, tree branches black against a blue sky. The boat seems a lot closer than when we left it. And it's listing.

Swain turns to me, a blush creeping up his face. "You're not going to write about this are you?"

The Tickler is beached.

There's no way I'll tease him, unable to read a tide chart myself. But Reynolds does as we stand in the estuary admiring the sunset and its reflection in the intertidal zone, killing a couple of hours until the tide comes up. Swain remembers the Fig Newtons he stowed and fetches them from the boat's cabin, silencing the ribbing as we nibble hungrily at the cookies.

While we wait, we have plenty of time to stroll over and examine the arching stone fish trap – the "whale's mouth" – now fully exposed, and to talk of adventures on the Central Coast. Reynolds once found himself with a doctoral student named Morgan Hocking, dragging garbage bags full of dead salmon through the forest and up mountainsides. The idea was to mimic how bears, wolves and other creatures drag their nutrient-rich prey through the understory, sprinkling food around for the green sedentary species lurking in the forest – the plants. Somehow, they dodged getting eaten by grizzly bears.

"Surprisingly, I haven't won a Darwin Award – yet," Reynolds says, referring to the honour bestowed posthumously on people who take themselves out of the gene pool through flamboyantly fatal acts of stupidity. Between them, Reynolds and Hocking would haul five garbage bags at a time – bags with a tendency to drip fluids. They'd struggle through dense tangles of ferns, devil's club and berry bushes, over moss-covered logs and up slippery hillsides, lugging the dead weight of dead salmon. When they came to a big log, one would station himself on the other side and the other would hand over the bags. They were strictly equitable, taking turns holding three or two bags. Every now and then, they'd stop, panting and sweaty, their pungent load at their feet, and silently question their own judgment. "You have to stop sometimes and wonder what it is you're doing exactly."

In this case, the half-baked methodology was part of a major study that Reynolds and Hocking had launched in the early 2000s with a team of about ten researchers, academics and members of the Heiltsuk First Nation, attempting to quantify to what extent salmon feed the ecosystem. The idea was to count the salmon returning to 50 streams, then survey the streams – a laborious, tedious and time-consuming process – and inventory plants in the watershed. Finally, they would take plant samples back to the lab to search for the signature salmon scribble on their consumers, nitrogen-15 (N15). Somehow, in the effort, Reynolds and Hocking talked themselves into dragging those garbage bags of salmon through the forest, which might well have won them the Darwin Award.

We muse about the salmon jostling for position, squirting out eggs or sperm, then being dragged, eaten and excreted by larger animals for the benefit of the whole forest system. It's a bit like having one very generous benefactor continually funding certain charitable organizations, the currency being not dollars but nitrogen.

It still feels odd to Reynolds that the Central Coast is nitrogen-limited. "You wouldn't guess it to look at how lush the growth is," Reynolds says as we walk the rocky beach, pausing to listen to a bird call. It's the high-pitched "roller-coaster" call of the Pacific wren, Reynolds says, the most common year-round resident in these woods; he's been identifying birdsong for us all day. "But that lushness is because they're sucking up all the nutrients they can get," he says. "A lot of these plants can grow year round." To sustain that kind of growth, he says, they deplete the soil of nitrogen. Some whole, dead salmon tumble back into the estuary and the sea without contributing much in the way of nutrients to land mammals and plants, he says, but a remarkable amount of the salmon corpses stay. He waves to the trees in front of us, most standing, some leaning, while others are deadwood stretching across the water, hovering above spawning pinks. "Sometimes, these logs or branches look like clotheslines with dead salmon hanging from them. It's incredible. And obviously that's going to keep the nutrients around a lot longer than if they're just going straight down some steep-sided canyon with very little obstruction."

Though it didn't win a Darwin Award, the garbage-bag study eventually showed that high nutrient levels distinguished the land around salmon-spawning streams from the land around streams with no salmon. Most of the time, as far as 35 metres from flowing waters, it was clear from the nitrogen which streams were spawning salmon. Those nitrogen levels, Reynolds found, determined which plants would thrive. Nitrogen-hogging salmonberry – which has salmon-coloured flowers – abounds in salmon-spawning watersheds. So does stink currant, but not wild blueberries. Counterintuitively, nonsalmon streams host more plant diversity than salmon streams.

Another study, running at the same time as the Reynolds and Hocking study, found an increase in bird diversity and density during the summer around streams with large salmon runs.[41] And yet another study, narrowing the focus to Pacific wrens, is finding that those nesting on salmon streams get a population boost.[42] Reynolds likes to think about what these connections mean beyond all the charts and graphs. If salmon affect the density of Pacific wrens, then they are fiddling with the sounds of the forest, you might say keeping the forest, literally, in tune. Without the salmon, the unmistakable melodious trills and operatic acrobatics of this tiny biological piccolo would decrescendo, perhaps to pianissimo.

The sea laps into the stone trap. It's getting chilly, but we're standing as though posed on a travel-magazine cover; the ocean reflects streaks of purple, orange and blue from the disappearing sun. *The Tickler*, a dark silhouette on the rocky beach, has perked up, water slapping its sides.

We wait a while longer before clambering aboard. Swain flips a switch on the control panel and *The Tickler* groans and sputters, deciding finally, with a shrug of irritation, to cooperate. The sky is black as we putter down the coast. The stars dance overhead, a child's mobile of celestial phosphorescence. We arrive back at the former cannery around 10 p.m. and after a quick meal, scatter to our beds; tomorrow we'll start early.

The next morning around 7:30, lunches and extra clothes packed, we head to the boat. Jeanette Bruce takes the wheel this time, a diminutive figure in the wheelhouse. A push of the button and the familiar sound of a moaning old man, followed by a cough, tells us *The Tickler* will start, just give him a few seconds. (Boats might be traditionally female, but to me, *The Tickler* sports a beard, pipe and a jaunty captain's hat.) We're headed for Roscoe Inlet, another fjord on BC's mainland, about 45 minutes away; we have five or six streams to hit today. Rain clouds gather as we chug through Queen Charlotte Sound. But as we close in on Roscoe Bay, the sun shines and the water glitters. The bay reflects an upside-down world so real that flipping the image makes no difference; the cedar- and spruce-lined shore can hardly compete with its reflection. Snow clings to the mountains in the distance, but before us the shoreline is a green velvet shawl tinged with every shade of orange and yellow.

Leaves, rocks, stumps and logs all shout their

colours. Then Bruce cuts the engine, and like at Gullchucks, the sound of rushing water overwhelms everything, except for the occasional punctuating splash of a salmon jumping. We hop out, and Swain anchors the boat, giving us a wave and a shake of his head, mumbling something about never, ever beaching a boat again.

As we wade ashore, a chum salmon leaps up and over a log, right onto the beach. Oops. From the human perspective, it's weird to see an animal miscalculate – what did the chum see or not see? Smell or not smell? Was the sun in its eyes? Reynolds scoops it up, a spawning male, recognized as a chum by its vertical purple blotches and fierce, dog-like teeth. Those teeth are what give the species its nickname, dog salmon. Reynolds carries it back to the stream's mouth, holding its tail as it wriggles in the water. When he lets it go, the chum hurtles along the same course toward the beach. And leaps. This chum seems programmed for miscalculation, repeated mistakes, at least at this particular, crucial moment in his drive to reproduce. Certain teenage boys come to mind. Reynolds picks the fish up again and carries it further, into the middle of the stream. Bruce and I tell him he's subverting the course of nature, the

Darwinian imperative of evolution. It's human arrogance, hubris!

"Sure," he says, wiping his hands on the sleeves of his t-shirt tucked into his chest waders. "But around here it's really impossible to divorce humans from the system anyway. They are part of the system – completely – and they have been, we know, for over 10,000 years. Right?"

Swain rejoins us as we watch the chum swim upstream, glistening in the sun, to join the fray of a few hundred chums working to keep their gene pool alive. The moment these fish have waited for their whole lives – at least in the Darwinian sense – will come down to being quick enough, aggressive enough and wily enough to drive off female competitors from depositing eggs in the best spot or male competitors from fertilizing those eggs. Or it might come down to being rescued by a biologist after making a wrong turn – twice.

As we stroll upstream beside the churning salmon, we chat about the concept of ecological baselines. It's a tool used everywhere by natural-resource-management types: Choose a point in history in which an ecosystem was assumed to be in balance or sustainable, and you tailor programs to return to that point. Atmospheric scientists' and activists' rallying cry, for example, is 350 parts per million (ppm), the highest sustainable level of carbon dioxide for stopping greenhouse loading. (In the early 21st century, we're close to 400 ppm.) The point might be to say that ecological baselines can be a useful tool in a valiant effort to bend human practices toward ecological balance rather than merely blind monetary gain. But, in restoring ecosystems, Reynolds thinks that determining a baseline can be an arbitrary exercise.

Westerners are most comfortable with written records, and even looking back to pre-contact depends on choosing a point in time, deciding how to read the archaeological record and determining what, exactly, we're looking for. Compared with salmon, some resources inextricably linked with First Peoples pop up relatively recently. Cedar trees, for example, only appeared here about 5,000 years ago. If the baseline is drawn to even 7,000 years ago, then the cedar's place is minor and perhaps warrants no consideration in managing the ecosystem.

Still, in contrast to other parts of the continent, which witnessed numerous advances, retreats and replacements of populations, the line from the earliest settlers to present-day First Nations in the

Pacific Northwest is unbroken. From oral histories, it's fair to assume that no new arrivals pushed aside or replaced First Peoples until the 19th century. Yes, some New World colonizers moved on to people North and South America. But many stayed. Over time, the climate changed, the landscape changed and the culture changed, but the Pacific Northwest has been home to humans and salmon since the glaciers receded over 10,000 years ago.

"You don't have to look very hard to see that the people were here and catching a lot of fish for a long, long time," Reynolds says. "You have to wonder if they co-evolved," he says, referring to the Darwinian notion of separate species reciprocally affecting each other's evolution, something that happens with species that have close ecological relationships.[43] Acacia ants protect the bullhorn acacia, a native tree to Mexico and Central America, from preying insects. In return, the acacia provides ants with food and shelter. If salmon and people have been living on this Canadian coastline for over ten millennia, people could have influenced the salmon's natal streams and spawning periods, enhancing the survival prospects of both. Scientists already know that hatchery-raised salmon experience genetic changes within one generation.[44] "That's a lot of generations of fish to be taken by a lot of generations of people," Reynolds says. Think about it, around the same time that First Peoples in BC were establishing a relationship with salmon, agriculturists in Sumeria were busy fiddling with plant genetics.

Ancient fishers here caught all five species, but after European contact something changed – the value of each species in the hierarchy of taste. There is no question that a fresh-caught sockeye or chinook is tasty. But take away the modern market and people value a fish like chum for the same reason we disdain and pay so little for it – it's low in fat. Chum salmon's size – it's the second largest of the five species – and its relative leanness made it a perfect fish for smoking and keeping on the Central Coast.

It's a fact that made an impression on Elroy White, an SFU-trained archaeologist and member of the Heiltsuk First Nation, after he interviewed community Elders about fish traps on the Central Coast. White's 2006 study, covering about 40 of the 250 documented traps on the Central Coast, contradicted what everyone – including younger members of the Heiltsuk community – used to think, that sockeye topped the fish hierarchy. The

Elders would patiently reminisce about catching and smoking chum, and White would try to steer them back to talking about sockeye.

> I always brought it back to sockeye because of that bias, and I was trying to be unbiased, which was a contradiction. And that really upset them, because it was as if I was not listening to them but disrespecting them.

It took him a while to see past the cultural bias he grew up with as a boy – sockeye rules![45] No matter where you grow up these days, you're unlikely to find a fresh chum staring at you from behind glass in a grocery store. It's always in pieces or it's canned, or it's jerky in an airport gift shop. If you want a fresh chum filet, you have to own a rod and know how to use it and figure out where to go, or find someone who knows where to go to take you.

The four of us watch a female chum facing upstream as she digs a nest. She turns on her side, curving her body as she sweeps her tail into the stream bottom. Habitat and species determine how fish make nests, but generally speaking, when female salmon come to spawn, this balletic earth moving is what they do. In the wild, it's easier to find chum spawning than other species because at this most important moment of their lives, their colouring is fascinating. In general, within a salmon species, the further north a fish lives, the more striking its colouring during spawning, but chum also wear a complicated costume. This female, tastefully iridescent, wears a bold, dark stripe along the length of her body.

As her tail sways back and forth across the gravel, I search for a satellite male – a male in drag, a subordinate morphing his colour to resemble a female and attract less aggression. Satellites – the metrosexuals of the males – are stealthy. They have to outwit the competition, big alpha males, who flank the females, fighting off other males, ramming into them and biting them or locking jaws. A smart satellite will hang back, away from the alpha, then zip in to fertilize the eggs at the moment they're released before anyone can stop him.

Alpha males can swim in drag too, if need be. In the 1950s or 1960s, Canadian scientist Kees Groot placed male chum in an aquarium, allowing satellite and alpha males to form pairs. Then he attacked the alphas with a broomstick handle.

Within seconds, the alphas' bar patterns switched to the horizontal patterns of a female. Once Groot halted his broomstick attack, the alpha male would nip a satellite male and – presto – gain back his stripes. That act of aggression switched the stripes on the chum's jacket from a prissy horizontal to a manly vertical. In the 1970s, Groot told PhD student Steven Schroder, now a retired fisheries biologist, about his observations. During his own research on chum, Schroder confirmed the behaviour at the University of Washington's Big Beef Creek spawning channel.[46]

Females are combative too. They fight each other for choice spots by muscling out the competition and, waiting for the machos they prefer, they fight off the males they find less attractive – the metrosexuals who look a lot like them and who they misidentify as competitors for spawning sites. Schroder noticed that when courted by males with stripes or smaller than themselves, they delayed nest construction by about an hour.[47] A female wants an alpha, an Arnold Schwarzenegger, a prime male specimen with garish purple and black vertical stripes that enhance his bulk. Likewise, a feminine appearance – the slimming horizontal stripe and dainty dog teeth (the teeth that elongate upon spawning) – must attract males, signalling them to hold off on the aggression. This is why switching stripes helps the satellite males; as long as they don't flash a nervous smile, one might say, they can pass as female.

To find the undersea equivalent of P.D. Eastman's *The Best Nest*, a female noses along a stream's gravelly bottom, testing it; too much silt and sand will smother her eggs. She looks for relatively calm water, a spot hidden from other females. Depending on the species, she'll dig from two to seven nests, cone-shaped hollows about 30 centimetres deep. Studies on coho have shown that bigger females have the muscle and power to dig deeper nests.[48] Shallower nests are vulnerable to disturbances of the gravel, whether by humans, weather or marauding females. Early arrivals get to the best nest sites first, but later arrivals will dig up their competitors' nests. This is easier to do if the earlier arrival laid her eggs and died already, being no longer around to defend her nest – although Schroder noticed that his chum females tended to die when their fertilized eggs were relatively immune to disruption. As ruthless and stealthy as males, female chums are made for a dog-salmon-eat-dog-salmon world.

Dig a little deeper and the beauty of biological adaptation reveals ever more intricate patterns. Early arrivals tend to be older and bigger fish, which presumably can scour deeper nests. Among chum, that means that the most successful breeders are 5 or 6 years old. Most returning chum in a Kuskokwim River system study in Alaska were found to be 5 years old – elders of the clan. Most, by far, were early rather than later arrivals.[49] The strong, deep-digging, long-lived chums, in fine Darwinian fashion, get to pass on their genes, insurance that their descendants will carry on into the future.

Swain now wades along the stream's edge, peering down at the slow-moving water. Reynolds squats on the gravelly beach, turning his gaze skyward when he hears a bird. Bruce stands still, water lapping her boots, mesmerized by the underwater bustle. Fish are splashing around, their dorsal fins, backs and tails peeking out of the water, which is less than knee-deep. We've hardly moved up from the ocean shore; chum and pink salmon spawn in a stream's lower reaches. Swain points out a couple of pinks, the males' backs humped, camel-shaped though thinner. They develop the humps during spawning season, which is why they're known as "humpies."

The humps, of course, have a purpose. Pink males, mature at 2 years old, form nest groups around a female, their positions hierarchically organized from biggest to smallest. The biggest male – unmistakable because he has the biggest hump – hangs closest to the female, guarding access quite jealously, even from smaller-humped buddies of his own nest group. But he tends to act more aggressively toward fish outside the group. So while passing males without membership in the club rarely get a fin in the door, a small-humped pink of the outer circle of the nest group has a fair chance of success if he dashes in and throws sperm over the eggs.

The water rushes, wind ruffles the trees, birds call – a symphonic accompaniment to the age-old ritual of pink mating. We mosey on, peeking upstream.

A juvenile bald eagle flies overhead as we later board *The Tickler*. Swain turns the boat south so we can check out Rainbow, a stream that is refreshingly hard to find on a map. In such a connected world, it's weird to find no reference to it on Google Maps. The glassy surface of the sea is mesmerizing. The only ripples are the ones made by the boat as we motor along, the land crowded with trees and rock that bump right up against the ocean. Rain clouds gather. We pass islands, here and there a cedar trunk tilting outward like a figurehead for a stationary boat full of trees. Arriving at Rainbow, we only pause briefly, staying in the boat. No salmon here yet. Another stream, named Quartcha is next, Swain says, turning the boat around.

At the boggy outlet of Quartcha we find some juvenile coho. Swain and Reynolds pause to fiddle with the radios to make sure they work. Standing in a thicket of sedge, we unwrap turkey sandwiches for lunch, turning up collars and hoods as water droplets coalesce in the air, not quite falling as much as swirling around us. Then it's off to Ripley, a nonsalmon creek.

A waterfall, invisible though within earshot from Ripley Creek's mouth, stops salmon cold. Ripley is Reynolds's "control" stream, where he can test the idea that salmon feed plants. It's one of the areas through which he and Hocking dragged garbage bags full of salmon, leaving the carcasses to fertilize forest plants that normally don't get a taste of salmon. And, yes, it is grizzly bear territory, something of which we are well aware as we scramble over downed trees and boulders along a path barely visible through the brush. A bear had apparently ambled up the same path earlier, a bear that had eaten a lot of berries, maybe too many berries.

"I think this bear might have a gastrointestinal problem," says Swain, as we pass the fourth or fifth huge pile of scat, whole berries still visible. We trudge along a trail, careful to hold back whips of foliage designed to take out an eye, and climb over fallen logs until we arrive back along the stream, plunking through the water as the trail peters out. We are spared an encounter with a crampy, gassy grizzly.

From Ripley, it is off to Clatse, a stream that traditionally has one of the larger salmon returns. *The Tickler* chugs past a floating log, 11 gulls perched on it in a neat line, calm sentinels granting our passage. Reynolds points out a couple of dainty ones with stylish pink legs and black smudges behind their eyes. These are Bonaparte's gulls. The bigger ones, California gulls, have more typically gull-like plumage, with greenish, snot-coloured legs holding up stocky bodies. *The Tickler* slows down and we glide into the estuary of the Clatse River, a bigger watercourse than the others.

Swain drops us off and anchors the boat. We hike. It's so green. It seems as if green has a smell all its own, of freshwater, dirt and peat all at the same time. I like it. We head back to the shore after a short survey, walking to a point from which Swain can pick us up. While Swain paddles a kayak out to *The Tickler*, we stand on uneven ground, thick with sedges. Reynolds saw wolves here a couple of years ago. I'd like to see a wolf, standing on the shore, eating salmon as they do in this part of the world.

I forget to jot down how many fish are here today. Swain has dutifully recorded the salmon at each stream, a slice of an aquatic world that, dutifully, too, swims to our world each year. And there's another bizarre world of salmon that rests in libraries. The scientific literature on them is more than a measly world, it's a universe unto itself, ever expanding, racing out to infinity with no contraction in sight. Salmon feed an entire scientific knowledge system, as well as a coastal rainforest ecosystem. But the scientific ecosystem is a flawed human construct, knocked out of balance by competing perspectives, desires and needs. DFO, for example, doesn't monitor many of the streams on the Central Coast on an ongoing basis. Chum and pink salmon are, after all, a low-value species.

Scientists know that since the counting started in 1950, chum have visited at least 119 streams on the Central Coast (although the true number is much higher). Some streams have been counted most years, others perhaps only once. Size matters. A bigger stream gets counted. Smaller streams do not. Clatse, a productive stream, happens to be one that DFO monitors fairly consistently. This year 6,400 chum and 54,000 pinks will come back, and only five sockeye.

By only monitoring strong runs, DFO risks making small streams into what scientists call "ghost" streams – ignored, unenhanced and eventually depleted.[50] It's an odd policy, considering that on the Central Coast, small spawning streams dwarf the number of large streams. Small streams, then, probably deposit the most nutrients into the forest from the sea.

Reynolds's program, which runs in partnership with the Heiltsuk Integrated Resource Management Department, has led to more than twice as many streams being counted each year. They share the information with DFO, hoping that the data will somehow inform management decisions.

In a way, Reynolds did an end run around the salmon hierarchy by looking at what chum and pink feed into – an ecosystem that includes bears and wolves – and not the fish itself. By showing the relationship of the fish to the forest, he steps back from the narrow lens of commercial interests and looks at life in its entirety.

It's not that Reynolds is a lone ranger. He's just another working stiff, motivated by the kinds of values that drive most ecologists, hoping to help tune up the planet, so that we – and all species – can endure and prosper. Scientists need to publish, and to publish, they need funding, and in chum and pink on the Central Coast Reynolds has found a dowry, a fount of grants. Charles Darwin was lucky to be an English gentleman with cash to burn. He was open in his gratitude for that.[51] He might have still written *The Origin of Species*, but maybe he wouldn't have had to time to write his influential book on soil science: *The Formation of Vegetable Mould through the Action of Worms, with Observations on their Habits* (or *Worms*).

Chum are endowing lots of scientists now. They're naturally the most widely distributed salmon in the Pacific, and like pinks, they're heavily produced and released from Japanese hatcheries (the 2011 tsunami knocked out some of their infrastructure).[52] They're definitely becoming more desirable: wild and farmed chum production has been steady since 1980; the wholesale price for chum has risen dramatically since 2002.[53] Three chum fisheries in British Columbia – though none of them are on the Central Coast – were declared sustainable by the Marine Stewardship Council (MSC) in early 2013, reflecting and boosting the marketplace status of this hardy, strapping fish.[54] Still, in the hierarchy of salmon preferences, chum rate low. Scientifically, too, others hog the spotlight.

Maybe this is because the average person knows so little about salmon. The average British Columbian knows how to cook a salmon, but can they look at the flesh for a clue as to whether it's farmed and not wild? No. Do they know that the chum salmon's digestive system and internal organs differ considerably from other salmon, or that they eat loads of jellyfish, accounting for their lower fat content and milder taste? No. If we're Salmon People in the Pacific Northwest, we're pathetic Salmon People. Chum, which may have had a great year in 2012, are still ugly Victorian brides, only desirable with big dowries.

CHAPTER FOUR

The Biological Black Box

And the young biologists tearing off pieces of their subject, tatters of the life forms, like sharks tearing out hunks of a dead horse, looking at them, tossing them away. This is neither a good nor a bad method; it is simply the one of our time. We can look with longing back to Charles Darwin, staring into the water over the side of a sailing ship, but for us to attempt to imitate that procedure would be romantic and silly. To take a sailing boat, to fight tide and wind, to move 4 miles on a horse when we could take a plane, would be not only ridiculous but ineffective.

—JOHN STEINBECK,
The Log from the Sea of Cortez, 51

THE MAHOGANY DECK on the *Ricker* – officially the Canadian Coast Guard ship *W.E. Ricker* – is a treat for the feet. I could stand here forever, I think, as I follow a deckhand on board. The dark, gleaming boards feel soft yet sturdy under my steel-toed boots. They promise an easy ten days on open seas. Standing on a deck for hours, days and weeks, sailors say, is hard work, like pushing ceaselessly against a leg press. For deckhands and fishers, tropical hardwood is an uncommon luxury, warmer than metal, easier on the feet and legs. Even fishnets last longer when they slip across the smooth, hard surface of a mahogany deck.

I'm with the High Seas Salmon Group, headed by DFO scientist Marc Trudel. The group's job? To chase juvenile salmon around the Pacific Ocean and figure out how life is going for them. The *Ricker* will take us as close as I'll likely ever get to what biologists call the "Black Box period" of the salmon life cycle. The open ocean is where salmon spend most of their lives, anywhere from one to seven years, stuffing their faces with fish food, from tiny zooplankton to forage fish a human would savour.

Salmon are the Takeru Kobayashis, the competitive eaters, of the marine environment. Instead of hot dogs, Twinkies and meatballs, they're gobbling copepods, squid, shrimp and herring. They chow down as much as they possibly can before the bell dings and it's time to stop cold and head home for their once-in-a-lifetime prize, the chance to reproduce.

We leave from Canada's federal DFO Pacific Biological Station (DFO-PBS) in Nanaimo, a small city on the east coast of Vancouver Island, opposite Vancouver on the mainland. Trudel wastes no time in beginning his survey. As we pull out of dock at 9 a.m., heading north on an overcast November day in 2010, he's down in the hold, "the lab," making sure labels, pens, scale and other equipment are on hand to greet the caught fish. The cloud cover thins to reveal our geologic companion for the trip, the snow-capped Coast Mountains, a 1600-kilometre-long range running from the Fraser River to southeast Alaska.

A boisterous bark of laughter announces Trudel's presence in the mess, behind me. The French Canadian's somewhat tough-guy image – shaved head, goatee, barrel chest – belies a playful bear of a man. He has the disposition of a good doctor, the kind who would explain carefully and tenderly the contours of an X-ray image and the treatment choices implied. Medicine was actually a career choice he tossed aside after a fateful encounter 25 years ago when he fell in love with fish.

Trudel's current research sounds simple, as is so often the case with biologists studying animals in the natural world: Are the fish that grow fastest best equipped to survive in the ocean? If so, can DFO better predict salmon returns based on measurements taken when they leave the streams as smolts?

"I try and understand why salmon come back big in some years, and in some years they don't," he says, holding up his hands and shrugging. But the puzzle Trudel is fitting together is huge, with endless tiny pieces – an image of the relationship between ocean environment and salmon survival.

Trudel and his crew spend three to four weeks at a time at sea, three times a year, slavishly following a routine. Like salmon returning to natal streams, they visit the same 25 to 30 data points, at the same time of day. Their work is the kind that rarely catapults a scientist onto the cover of *Scientific American* – the seemingly drab data gathering of long-term environmental monitoring.

But by now, after four trips with salmon scientists, I perceive not only the wonder of their daily work in the wild – the satisfaction inherent in a quest to understand the stuff of our world – but the beauty in the drab charts and numbers. They echo our mundane relationship with the planet. They track our environment with the grace of growing precision, and they do so with a high purpose: to see if it will support us.

Sampling the ocean daily, the scientists on board the *Ricker* perform a well-choreographed dance to the tune of groaning cables, rattling chains, splashing water and the slap of fish on a steel table below deck. And, oh yes, bongos. Or, as the captain chimes in over a lunchtime explanation in the mess, the rhythm of the ship is, from dawn to dusk, tripartite : "CTD, bongos, fish. CTD, bongos, fish. CTD, bongos, fish."

About 65 kilometres north of Nanaimo, near Lasqueti Island in the Georgia Strait, the dance begins with the CTD array, the tool named precisely for what it does, which is to measure the ocean's conductivity, temperature and depth. The cue for the dance is a grinding sound, a winch lowering a collection of instruments including a Niskin bottle, a vessel about the size of a self-serve coffee urn, open at each end, with stoppers connected by an elastic cord. When the bottle is at an ocean depth that Trudel wants sampled, a weight travels the cable to hit a release mechanism attached to the elastic cord, closing both ends of the bottle to collect water. Violently. No commercial crossover here as a bathtub toy.

The Niskin bottle is one of many technological innovations that confirm the peculiar economy of marine biology, a market reliant on ocean engineers, leaps of ingenuity and solutions not always transferrable to daily problems. Where moon exploration gave us the hand-held, cordless mini-vacuum – the Dust Buster – ocean exploration, and inventor Shale Niskin, gave us Niskin bottles.[55]

Measuring conductivity is exactly what it sounds like: seeing how well the ocean, at a given point, conducts electricity. High conductivity means high salinity, low conductivity means low salinity. Salinity plus temperature (T) tells scientists something about seawater density, a major force driving ocean currents. (Denser, colder saline water sinks, while warmer, less saline water is buoyant, so the deeper a Niskin bottle goes, the higher the salinity should be.) Simple temperature readings also say something about warming or cooling trends. The result is a series of snapshots taken over time of the physical properties of the water column, at a particular time and place.

The Niskin bottle also brings back clues to the microscopic life of the ocean, and thus to the lives of salmon. At ten metres deep, where phytoplankton concentrate, chlorophyll levels in the water allow for a report on algal biomass. The more chlorophyll in the water, the more phytoplankton. The more phytoplankton in the water, the more zooplankton. And the more zooplankton, hopefully, the more fish.

The CTD, holding a full Niskin bottle, grinds back on board, and the "bongos" start – two connected cylinders rising from the mahogany deck, resembling a giant-sized pair of bongos. Mesh attached to each bongo flows, cone-shaped, to its own small, white bucket with screened holes.

The bongos, attached to a cable, swing out behind the *Ricker*, which tows them for ten minutes, collecting the Happy Meals of the juvenile salmon world: euphausiids, known as krill, and copepods, tiny crustaceans with a big impact.

Whether in fresh water or salt, swimming in deep ocean trenches or burrowed in streamside litter, copepods are the most numerous multicelled organisms of the watery world.[56] And as tempting as it is to make fun of a professional organization with the acronym WAC – World Association of Copepodologists – breathe a sigh of relief that these scientists have our backs. If copepods disappeared tomorrow, we would be in trouble. The woman on deck wearing a survival jacket, hard hat and rain pants while waiting for the bongos to haul up our zooplankton catch is not, however, a copepodologist. Mary Thiess is a statistician.

As she waits, she hoses the white buckets with salt water and detaches them. I follow her as she places them in bigger buckets and hauls them toward the ship's deckhouse and the closet-sized lab that opens onto the deck. Outside the lab, Thiess dumps a bucket's contents into a big canning jar laced with formalin and borax for preservation. The plankton will pickle until a technician gets around to identifying each species. As they cascade into the canning jar, the most visible creatures are the krill: big-eyed sea monkeys.

Through progressively smaller meshes, Thiess screens the buckets' contents three times. Cold water washes over her long, slender fingers. She keeps them limber between what she calls "data points" – the bongo hauls – by sitting in the mess, crocheting hats. The ocean is cold here year round. So is the air on this blustery fall day. Thiess's fingers, wrinkly from the cold water, scoop the catch from each screen into a separate plastic bag. She labels each bag with a Sharpie. I'm wondering how much data scientists lost before the invention of plastic and Sharpies, the remains of which will probably puzzle future archaeologists, who will unearth them with the same glee that archaeologists today experience when they come upon ancient weir stakes.

The life in the plastic baggies will be ground up and run through isotope analysis for carbon-13 (C13) and nitrogen-15 (N15) signatures. Carbon levels give hints as to where zooplankton, like copepods, do most of their dining. Do they prefer supper on the ocean floor, where they sink at night? Or do they favour brunching on planktonic algae closer to the surface? Nitrogen levels, in contrast, tell scientists where organisms rest in the food chain. Each 3.4/ml increase of nitrogen indicates a step up on the food chain, a position that can vary for omnivores like copepods (and humans) over the course of a lifetime. A fluctuation of nitrogen levels in a species can tell researchers a good bit about changes in the food web, whether changes in the environment have led to changes in predator/prey relationships.

While Thiess screens the take from the bongos, the Coast Guard crew – mostly former fishers from the east and west coasts who can no longer make a living as their parents and grandparents did before them – ready the trawl net. For half an hour at varying depths – surface, 15 metres and 30 metres – the *Ricker* tows a net as big as a basketball court. The goal is to net juvenile salmon. In the Strait of Georgia, only about 28 kilometres wide, lots of fish live within a narrow habitat 15 to 30 metres below the surface. Think urban density, as in Montreal, compared with the west side of Vancouver Island, where the fish can spread out in an underwater version of Calgary sprawl.

For the finale of the dance – CTD, bongos, fish! – Trudel and Yeongha Jung, a South Korean-born fisheries technician as easy-going and quick to laugh as Trudel, count, weigh and sometimes dissect the fish they catch in the trawl net. It's my favourite part of the routine. Depending on the time, day and net depth, they may come up with hardly any fish or they may haul in a whole teeming city of them, as diverse as Toronto. Then I'm a 5-year-old watching her mother bake. I make it my mission to see, touch, smell – "Can I help?" – and soon I am, if not an expert, at least an apprentice. I learn to identify the juveniles of the five salmon species, to equate the smell of fresh cucumber with whitebait smelt, to understand – painfully – that plunging one's hand into a plastic laundry basket full of deceptively harmless forage fish is a bad idea and to see spiny lumpsuckers as damn cute. It's fun mucking about with all kinds of fish, all of them are fascinating.

So why do salmon get all the glory? And why do some salmon get more glory than others? Why do sockeye and chinook vie for the top spot, followed by coho? Why are chum and pink in a distant tie for third? Right, well, it's money, for the most part. In pure commercial terms, sockeye is the billion-dollar industry. They entice sport fishers with their expensive gear, they flood grocery stores in fall and they are canned and exported mostly to the United Kingdom and Australia.

It's partly a matter of abundance and partly a matter of which species the fisheries choose to nurture.. There are more sockeye in the world than chinook and coho. But it's also a matter of preferences. In surveys, consumers say they prefer wild sockeye or chinook to other salmon.[57] Taste, of course, is learned and bound with fashion; lobsters were once poor-people food. Scientists, with their empirical habits and skepticism of advertising, might be expected to diverge, and in my sampling, they often do. When I ask them which is their favourite salmon, most don't mention sockeye. They practically drool, though, when they put "freshly caught chinook" and "barbecue" in the same sentence.

Late in the afternoon, Trudel and I repair to the mess to drink tea and eat cookies, and when I ask him which salmon species is his favourite to eat, he surprises me.

"Coho," he says. "They're more consistent, taste-wise. Sockeye: depends on the run." Since most sockeye depend on lakes for their early life stage and spawn near and far – from 100 to 1,200 kilometres from the ocean – their fat content varies more than coho or other fish, like chum and pink, that spawn closer to the mouth of a stream. Some fish have further to go than others. "The fish that have long migrations have to be fat. Those are the tasty ones. Or the tastiest. The ones that have a short migration usually tend to have a whole lot less fat and they're not as tasty. They're still very good, don't get me wrong. But in terms of consistency of taste and quality: coho."

By 5 p.m. we are nearing Cape Mudge on Quadra Island, home of the Wei Wai Kum First Nation, part of the Laich-Kwil-Tach, the same group that Christine Roberts and Randy Dingwall belong to, the two archaeological crew members with whom I slogged through the Philips Arm estuary. An anthropologist recorded their creation story about a hundred years ago.[58] Such records are always suspect. It's plausible the teller purposely misled the anthropologist, or the translator stumbled. Or, like all translations, the rendering was only approximate, or the anthropologist misled himself because he tried to squeeze an unfamiliar narrative into a beginning, middle and end, the narrative arc of his own culture. But no matter how our version of the story was created, it is still part of the First Nation's cultural history and the best part is when Raven finds himself at war with the Salmon People, who in turn summon the other Fish People, from smelts to killer whales. After some kind of mash up, Raven throws salmon, eulachon and others from his canoe to different rivers where the fish people ultimately make their homes. These "fish" – whether fish or sea mammals – have homes and personhood. Their origin story – in conflict with Raven – is a metaphor for what in scientific terms is an animal's "life history," the strategies it uses to stay alive long enough to reproduce. The story explains why sockeye fry prefer lakes, chinook fry vary in how much time they spend in fresh water and coho fry prefer a stream's back eddies.[59]

The 2-year-old coho I see on board the *Ricker* have another year to go before their mighty journey to reproduce; coho are on a three-year spawning cycle. When Trudel talks about coho, he always contrasts them to other salmon species, similar to the way we explain different human cultures in relation to each other. Coho and chinook roam the ocean eating big prey – anchovy, herring, sardine, pollock, sand lance and smelt – while other salmon species fill up on smaller organisms. A just-caught juvenile coho smells fresh, while a chinook smells fishy. Chinook have black gums, while coho have white gums and a black tongue. To contrast one salmon species with another helps scientists figure out what factors will contribute to the survival of one species or another and throws the differences among them into high relief.

That evening we steam right past Quadra Island; this is a sea voyage with no plan on docking until we hit BC mainland's most northern coastal city, Prince Rupert. It's dark, the fishing is done and everyone arrives to the mess at dinnertime, no taking the cooks for granted. Three square meals a day and snacks. When we're served sockeye, I ask the scientists whether it's farmed or wild. "Wild," Thiess answers. Trudel points to the myomeres, the long thin lines running the length of the red flesh. In farmed fish, there are bigger spaces between the lines and a layer of fat. "Plus," Trudel says, touching his fork to a pickle, "it would be this colour."

The big screen television in the mess is on almost continuously, the screen filled with race cars, or a game show, a television series or movies. Will Smith's face fills the screen and he yells, "I can save you! I can save you!" If a zombie apocalypse was to suddenly occur, à la the movie on the screen *I am Legend*, sailing the sea would be, I think, a stroke of luck. When I head below to my quarters, bunk beds, a closet, a set of drawers, a desk and a washroom connecting to Thiess's room, it feels safe and cozy. Someone is always on the bridge, watching and ready.

By 8 a.m. the next morning, the *Ricker* is plowing through the shimmering waters around Calvert Island, past Vancouver Island and about 60 kilometres south of Bella Bella. For the rest of the journey we'll traverse the territories of the Heiltsuk, Wuikinuxv, Nuxalk, Kitasoo/Xai'xais and Haida. By mid-afternoon we pass Namu on the mainland, a significant archaeological site and an old cannery town at the junction of Fitz Hugh Sound and Burke Channel.

When not peering over shoulders into the business of the ship, I wander to the bridge to look through binoculars, the landscape slipping by in a peaceful monotony of trees, snow-capped mountains, sheer rock faces and waterfalls. The area was probably habitable as early as 13,500 years ago when the sea was lower. What today is Queen Charlotte Sound, to the north and west, was once a coastal plain of lakes and streams where a living could be made. The earliest archaeological evidence dates human settlements here to about 11,000 years ago; at Namu archaeologists have found salmon bones mixed in with stone tools and other artifacts.

When the deckhands haul up the trawl net, I hurry to the lab, although I don't see much action within the net. Having chugged around these waters in a small boat, I feel a little closer to these fish.

The fishers deliver the fish through a trap door from the deck to the hold and into a laundry basket. At this moment only three chinook and two chum loll in the basket. Jung had already removed the bycatch, a lumpsucker and sturgeon poacher. They recover together in a bucket. Trudel will eventually write a paper on fish assemblages they have caught over the years.

As I watch the hauls come in, I realize it's a lot of smelt, herring and other forage fish, as well as squid and jellyfish. One net dumped 29 kilograms of white bait smelt into the hold, their wriggling bodies and big-eyed stares a testament to a desire to live. They taste good over a fire after the heads are hacked off and the guts removed. Or so I'm told. After handling thousands of smelts, I still have no idea what they taste like. As a coastal resident, why haven't I eaten them or herring? Later we'll find a pollock in another load of smelt. Trudel, a numbers lover who took statistics in university instead of another biology class, does a quick calculation after weighing and counting the fish and says there was a one in 2,361 chance of finding a pollock in that load of smelt. We sift through the slivers of smelt looking for more pollock but only find a few herring. Looking for salmon is a priority, but the little fish fascinate me.

We place the salmon on trays to be labelled, measured and weighed. Jung punches a hole in a chinook's gill for DNA sampling, using tweezers to place the bit of flesh into a plastic vial with a label. He shows me how to extract the otoliths, the ear bones. Wearing surgical gloves, Jung snaps back a salmon's head from the gills, grabs scissors in one hand and snips into the head near the brain. He puts down the scissors, grabs a long set of tweezers and plucks out the otoliths. A bit of blood stains his gloves. If you hit the wrong place – as I do over and over again – and blood spurts, it makes it harder to find the shiny white otoliths But a week later I'm good enough to look forward to dinner parties, a big chinook resting on a shiny platter, friends gathered around the table, admiring how I snap the fish's head, snip into the brain and pluck out the prize, the otoliths, with the long tweezers that happen to be on the table. A fight breaks out, who gets to go home with the prize? If the otoliths are big enough, they make nice earrings or cufflinks.

The ear bones I remove from the fish are about the size of little white rice grains. Otoliths record a fish's age and the number of days it has spent in the marine environment. The little bones develop circular rings each day, although as the fish ages, the rings are so small, they vanish. "The more space, the faster it's growing," Trudel explains, as I snip at a chinook head. Holding the slippery fish in one hand is tricky, especially with the head almost dangling off. "Like a tree, but on a daily basis." Rings with a different rhythm than growth rings mark environmental conditions, and checks on the otoliths that look dark under a microscope mark eventful moments: hatching, entering fresh water, entering the ocean.

Through their ear bones, then, the fish can tell us how fast they've grown, at least in their first few months at sea. Trudel has been wanting, for years, to use the ear-bone rings to measure growth. But like so many things, funding sets the agenda. "Finally we're starting to do this," he says.

The boat chugs further north. Then at the right transect (remember, Trudel checks the same locations year after year), the dance begins again: CTD, bongos, fish. The next haul nets six juvenile chum, four pinks and one sockeye. Down they go through the hatch, into the fish lab where Trudel places the fish on a tray. He lifts the gills of each salmon to expose the gill rakers, the sharp spines lining the gill opening.

The pink salmon's gill rakers are fine, like a fine, black comb that tucks into a guy's back pocket. The chum's gill rakers are stubbier, spaced wider, more pliant – as if the comb melted away a bit and lost a few of its tines. The sockeye's are fine too, but long, like blades of red-stained grass or sedge. Juvenile salmon are particulate feeders – their jaws act like a vacuum, sucking prey right out of the water and into their mouths – so the difference in gill rakers puzzles Trudel, except maybe for the sockeye, a salmon that may filter feed, too, their fine, long gill rakers straining aquatic food morsels from the water column.

Describing salmon – assigning it a family, genus and species used in Western science – began in the late 18th century. The only known salmon for most of the 18th century was the Atlantic species (*Salmo salar*) native to the North Atlantic and Arctic Ocean above Western Europe. A German taxonomist, physician Johan Julius Walbaum, who worked for the Russian Imperial Court, was the first to scientifically describe Pacific salmon, in 1792, though he had never seen one.[60] It was not unusual for taxonomists at the time to identify varieties and species, either by reading descriptions and looking at drawings or by studying dead specimens sent from explorers or other naturalists. (Walbaum was also the first scientist to wear sheep intestine gloves to prevent the spread of infection.)[61] Darwin saw this as a problem in *The Origin of Species*.[62] Vague descriptions scribbled by poorly trained scientists or specimens plucked from an environment clear across the world simply wouldn't do. Almost two centuries after Darwin set sail on *The Beagle*, Trudel has a full toolbox for identifying his catch. When he finds a fish in the trawl-net haul that he can't describe, he reaches for a worn paperback book, *Pacific Fishes of Canada*. Or, more often, experience trumps the book, and he dashes to the upper deck to ask one of the former fishers to come down and have a look. Genetics, too, though, have changed taxonomy, reaching far into the past to map connections.

Jung has filled "box number three" with chinook DNA samples, bringing the number of DNA samples for that fish to 300 since this season's research began about a month ago. Now he grabs a hole puncher and cuts a circular DNA sample from the sockeye, like he did with the chinook. They'll send these, along with bits of coho, to a lab in Vancouver for analysis. No such deep study is wasted on chum or pink – each DNA sample costs Trudel $20.

Since 1998, scientists with the High Seas

Salmon Group have analyzed more than 6,000 samples of sockeye and more than 6,000 of chinook. Scientists match the DNA to populations, which gives them a sense of how a given salmon population fares. To make a match, a salmon population has to be tagged before they head out to sea. This is time-consuming and expensive. Even for sockeye, a commercially valuable fish, only a few populations have been tagged: the Cultus Lake sockeye from the Fraser River system and two others from the Columbia River system, the Wenatchee and Redfish.

Trudel's analysis confirms the tagged fishes' identities, adding to a baseline sequence of DNA for a fish population. Gathered over years for comparison, DNA tells scientists something about the genetic diversity of a population and its structure. But first, a baseline needs to be created. Cultus, Wenatchee and Redfish sockeye have a good baseline and their DNA matches against the tag 100 per cent of the time, an effort that took years and thousands of fish. For example, in 2011, the accuracy for Redfish sockeye only reached 85 per cent, so three out of 20 fish were classified incorrectly until the baseline improved. Most salmon populations have no DNA baseline at all.

Though Trudel is a federal government fisheries biologist, like all research scientists, he scrounges for funding, putting together a mish-mash of cash from the Natural Science and Engineering Research Council (NSERC), the Bonneville Dam Corporation and whatever other funders will climb on board. He was far from immune to the economic meltdown. In 2008, the same year the housing bubble popped, hammering global financial markets, Trudel had applied for funding for a dream project. He's been developing a model to better assess how many coho, chinook and sockeye salmon return to major rivers in any given year, an equation for a growth rate that would serve as a forecast. If the salmon he analyzes grow at a particular rate, Trudel can tell the likelihood of a successful return to their natal streams.

"My first model was for coho," Trudel says. We have repaired to a small office off the fish lab. Trudel sits behind a desk. We're between fishing points, the sound of the engine is loud in here as we chug along, the slight vibration a constant though imperceptible buzz. "If you can figure out how fast they grow in a finite amount of time, you can predict coho returns," he says. If you can do that, he says, understanding the bounty is merely a matter of counting the smolts that leave the rivers – again, another time-consuming task on a coast rich with salmon streams. The only counting of coho smolts that goes on is at Carnation Creek, a wild stream that experienced a huge decline in salmon in the 1990s, and Robertson Creek, a hatchery, both of which are on Vancouver Island close to where Carol Schmitt babies her chinook. The territory salmon traverse in the ocean may be immense, but the BC salmon world of humans is small. With a little coordination from the fisheries – counting smolts as they leave – an ability to predict returns and manage salmon sustainably would improve by leaps and bounds.

To create his model, Trudel plotted seven data points from 1998 to 2004 on a graph. On the x-axis was coho growth, on the y-axis was coho survival. The faster a fish grew, the more likely it would survive, swimming through the winter, inhaling food. In 2005 Trudel noted that the salmon weren't growing much while at sea – they had fewer large, fat copepods to feed on and were swimming in warmer water temperatures. Warmer water temperatures also brought warm water predators to the coast. The 2005 juveniles that Trudel caught would head back to their natal rivers in 2006, a year for which he forecast poor returns. He was right.

"I like to use the phrase, we're going outside the known universe – something I picked up from the captain of the *Ricker*," Trudel says, his voice booming over the engine's din. "I thought that was a powerful sentence, because the model is constrained by what we see. We don't know what the upper limit is [for coho]. In 2005, we were probably close to the lower limit."

That lower limit was about 0.5 per cent of the fish in those streams surviving that year. Although, as Trudel points out, the lowest could be 0 per cent, which would mean a wiped-out run.

The *Ricker* cruises as far north as the top of Haida Gwaii in the Dixon Entrance, just south of Alaska. Fog wreaths the ship on the morning we turn back south. Our airborne companions, seagulls, materialize next to us then fade, replaced, endlessly, by their brothers. The CTD deploys as the day dawns, a reprise, the sound in time with the seagulls' wing beats. We catch no fish in the first haul. No specific reason is given; fish are fickle, chasing a meal, evading predators.

The science on board the *Ricker*, however precise, yields a view as murky as the tip of Haida

Gwaii. Every once in a while a sliver of information materializes – salmon growing big, according to Trudel's calculation, will live to spawn – but the scientist is also the first to acknowledge that he's looking through a peephole, catching only a tiny glimpse of the whole picture. Studying ocean-going salmon will always be opaque compared to studying them in rivers and streams. We can watch salmon race up streams and rivers; watch them spawn, die and emerge from eggs; and we can study their habitat, even without being there since webcams have opened the wilds to scientists and citizens alike. To spy on the ocean in the same way seems beyond even future technology.

On board the *Ricker*, I get the sense that the constraints on research are like a low-level, chronic infection, always there, festering in the background. Some days, weeks, years, scientists feel great. They have the funding and resources for innovative work, leading to breakthroughs, resulting in healthy changes in a management system. Then they're struck with budget-itis and shifting political winds. (Or, as humans, they can also be struck by another malady – bored-done-tired-of-chasing-funds.) Trudel has the passion but no key to a bank vault and his career has coincided with a federal government allergic to scientific research. The tragedy of a profit-driven world is that knowledge has a price tag, when, like salmon and other shared resources, it's priceless. After ten days at sea I can look at juvenile salmon placed on a tray and identify the species. Ten days. What does someone know after a lifetime? How can we put a price on that?

About a year and a half later, in May 2012, when the chance comes to get a little closer to coho, the most elusive of the salmon species, I take it, climbing on board a much smaller boat than the *Ricker* for a shorter journey – to Namu from nearby Calvert Island. By now, the days are long, it's warm and most of the juvenile coho have fled the streams around Namu. Will Atlas is another passionate salmon scientist, about 15 years younger than Trudel. Joking that he's part fish, Atlas looks for his aquatic brethren over the side of the boat as we approach a dock at Namu. Atlas, his blond hair in a ponytail, a ball cap on his head, asks the two university students he brought with him from the Hakai Beach Institute, a privately funded research centre on Calvert Island that holds summer field courses, to ready the equipment: coolers, nets, plastic bags, Sharpies.

He plunges a hand-held net from the side of the Hakai boat in search of juvenile salmon of any species. He hauls up a dripping net full of fish, passes them off to the rest of us and we place them in the sea-filled coolers. Atlas systematically grabs them one by one and with a flick of his finger to the fish's head, it dies immediately. We package the dead fish into what's called "whirlpacks" – fancy-looking Ziploc-type bags – for later lab analysis. The tissue samples will reveal any diseases in the fish. But scientists are also obsessed with what salmon eat, and Atlas wants to know what these fish, mostly sockeye, but coho, too, scarf on during their way out to sea. These coho meandering around Namu will hug the coast a little tighter during the marine phase of their lives, closer to their natal streams than sockeye, avoiding the deeper, wilder, open ocean that might offer a bountiful buffet or not much at all.

"This place is a good place to make a living for a juvenile fish," Atlas says. Where rivers and ocean meet, life teems and finding food is easy. Atlas selects a fish from the cooler, places it on his palm and turns it over, revealing the black spots on its back that mark it as a coho, not a sockeye. He points out its long anal and dorsal fins, another feature that distinguishes the coho from its cousins, probably an adaptation to the slow-moving water cohos prefer when they are young and avoiding predators in their natal streams.

We tie up at a floating dock. Namu, a cluster of dilapidated buildings sitting on aging docks and climbing the hillside around a cove, was traditionally known to Indigenous people as Na'wamu or Ma'awas. Europeans turned it into a cannery town in the late 1890s, and the population swelled to thousands of workers. The settlement hummed for a while, with a nursing station, store, restaurant, post office, school, dormitories and machine shop with a forge. But all that was left behind when the last operator, BC Packers, closed the doors in 1985 and walked away.[63]

Now it's all dissolving back into the land and sea. Grey two-by-fours litter shoreline rip-rap, pilings are bereft of a dock. An old coastal freighter, the *MV Chilcotin Princess*, bleeding rust, leans against the rotting piers supporting a former warehouse. The liveliest sights are the trees and shrubs reclaiming houses and streets.

Once the salmon were gone, so were the people.

Well, not quite. Three elderly people still make a living at Namu. The floating dock we tie up to belongs to Pete and Rene Darwin – I forget to ask if they're related to *that* Darwin – a couple in their mid-60s who assembled it, along with several small buildings it supports, built from salvaged logs and lumber recycled from the town's collapsing buildings. They also set up a party float with picnic tables, a fire pit and a gift shop for visiting boaters in summer. Overnighters pay a 75¢/ft. docking fee – the price hadn't changed in eight years. Along with a childhood friend, Theresa May, they also tend a greenhouse, where they grow fruit and vegetables, peaches and kiwi, spinach and chives. And in the fall, they fish for sea cucumbers and urchins.

The trio are Namu's "caretakers," a nebulous job description, with little in the way of either specific duties or pay. Legal title to this modern ruin and some 100 acres around it is held by Namu Properties, owned by Langley businessman David Milne. It's been for sale for over a decade, but, as Milne says, "It's a complicated property" to unload.

We walk through the ghost town, past alder saplings coaxed into shapes – a heart in one instance. The faint smell of fish oil lingers amidst the intact buildings, some rented by local fishing lodges to overwinter boats. Poking your head into the store is like peering back in time. It's still "stocked" with magazines, engine v-belts, shelves of paint, Model T spark plugs, fluorescent light tubes and pallets of tartar sauce, mayonnaise and jam – the last a sweet lure for marauding martens, resident weasels smart enough to break open the jars and feast. The store had a liquor licence until 1995.[64]

We head across a broken boardwalk, shrouded in moss and overhanging tree limbs, the sounds of rushing water getting louder until we reach Namu Lake, its clear, cold water heading down a stream toward shore. Tucked into the bushes a few metres from the boardwalk we find the canoes a Heiltsuk band member said we could use, saving us the effort of carrying our own from the boat. Climbing into two canoes, the six of us push off and set a quick pace, gliding across the placid lake surface. We paddle about five kilometres. A student, Wanli Ou, sits in the middle of the canoe, on the bottom, while I paddle from the front seat. Rod Wargo, a Hakai employee and former fishing guide, paddles and steers the back. Rod originally worked at Hakai as a guide when it was a fishing lodge. He remembers taking Eric Peterson out fishing when the scientist-turned-entrepreneur visited the lodge, checking out its potential as a scientific institute. It passed muster. Peterson eventually bought the lodge in 2009 and kept Rod on. He's a good mechanic and knows the area well, both valuable traits to visiting scientists. Wanli Ou, with her laid-back demeanour and beat-up outdoor gear is pure West Coast, so I'm surprised when she tells me she's from Singapore. Ou has found her spiritual home on coastal BC.

We arrive at the mouth of a cold, clear, salmon-spawning stream, one of scores that tumble into Namu Lake. We swing our legs over the gunwale into the water, sinking into the muck. We haul the canoes onto land. An air horn hangs from the branch of a spruce tree, a handy noisemaker to scare off bears. We gather backpacks and clipboards, tape measures and quadrangles and other measuring paraphernalia and follow Atlas up a stream, jumping a little every time he blasts out, "Hey bear!"

Grizzlies don't like surprises.

Our purpose is to assess the salmon habitat, measuring, among other things, the tree canopy, the stream width and depth and the size of rocks on the stream bed. It's tedious work. And like all things in science, Atlas is taking this moment to build on the work of another scientist, who took on the monumental task of surveying at least 50 streams in the area in the same manner. That was Reynolds's protege Morgan Hocking, when he wanted to figure out what plant species the salmon streams fed.

After a couple of hours of standing in icy cold water, looking up to calculate the tree canopy percentage at different points, spooling out the tape measure from one side of the stream to the other, plunging our hands into the water to lift rocks and calculate the percentage of plant life attached to stone, we've advanced maybe 100 metres, passing through slender rays of sun, little of the warmth penetrating our boots to our feet. Eventually Atlas says we've done enough and we make our way back toward the stream mouth and stand amongst soggy sedges, eating our sandwiches, warming up under a clear blue sky.

Atlas saunters over to the stream bank and beckons to me. He crouches, pointing to flashes of movement near the exposed roots of a tree amidst a tumble of leaves, ferns, sticks and moss-covered tree trunks. "Coho like to hang out in these eddies. The water flows more slowly and they have some cover," he says, edging close. And there they are, a handful of little fish darting among the debris. I hadn't paid much attention to coho fry when in the field with Reynolds; we had focused on pinks and chums, the coho few and far between anyway. I hadn't realized how much harder it would be to see coho. They revel in the clutter of the streamside debris. It's a peaceful, elfin environment, spruce, cedar and hemlock trees dripping needles into the flowing water, the burbling of the creek playful, the sound less predictable than the sound of a running toilet I associate with urban water features plumbed into groomed gardens.

Coho prefer freshwater streams with low gradients and slow-moving water, a peaceful environment, yes, but it leaves them vulnerable to dry summers and autumns. Without much precipitation, their homes – smaller and shallower than those of the lake-dwelling sockeye – dry out first. It's risky. But their hidden niche at least offers places to hide amongst woody debris and overhanging plant life from bigger fish and birds, plus less competition from other salmon species. Coho stuff their faces for about a year before they migrate to sea. One of British Columbia's most beloved naturalists, fly fisher and writer Roderick Haig-Brown, writing in 1959, called them the most voracious of the five species:

> They come down from their streams after a year of aggressive fresh-water feeding – they are most aggressive of all the salmon and trouts in their pre-migrant stage – and quickly put their activity and determination to good use amid the new abundance; by February or March of the following year they weigh almost two pounds; by the end of May probably three pounds; by July five or six pounds; and in October they are fully mature fish of eight or ten pounds, ready to go up the rivers to spawn and die.[65]

Atlas, a freelance biologist, has spent the past few summers working on the coast with organizations like the Hakai Beach Institute and the First Nations, all working collectively to draw a portrait of the Central Coast through science. A lot of attention is lavished on salmon, but considering their role in the ecosystem, it's understandable. It's much easier to explain to people why this animal is important – if nothing else, they taste good. Again the knowledge Atlas adds is like a strand in a web, not the kind of research that makes a scientist famous.

Atlas and I wander along the stream toward the lake, the stream a few metres across, as deep as my calves, coursing through a mudflat of sedges. Deep paw marks, visible between strands of the edgy green stalks, tell us a bear visited recently. The stream seems too small to nourish as much life as it does and it's easy to see how years ago, when loggers sliced through the coastal forests, right down to a stream bank, they failed to understand its importance. Only the locals would understand.

Coho are often associated with small streams like this one – a "ghost" stream – a creek with no formal name. For most of us, a stream's remote location and thick vegetation mutes its importance. We can't get to them easily, so they don't exist for us. A diverse population of mighty bears, wolves and eagles know them, though, and the quieter, cryptic, rough-skinned newts, banana slugs and lichen moths, the creatures that communicate without words. Scientists are driven to experience such worlds – worlds that communicate through different channels. But with their numbers, graphs and language, scientists sometimes lose track of how these worlds affect them, change them and steer them. Or at least they don't talk about it much.

Fisheries biologists start thinking like fish.

Ecologists start thinking like animals and plants. Archaeologists and anthropologists weave in a human element: where did we fit in, before we planted ourselves so firmly outside the natural realm? All these scientists try to translate for the rest of us, but they're translating from a language so distant that much of the meaning is inevitably lost. Only by experiencing the worlds of animals and plants – as explorers, as gardeners, as fishers, as hunters, as foragers, as scientists and naturalists – can we understand them. But most of us spend our days and nights indoors, boxed off from those worlds.

Why are nature shows endlessly consumed on the Internet and television? It's a proxy for living in those worlds, for thinking like a shark, or an elephant, or a grizzly bear.

CHAPTER FIVE

Life, Anthropology and Everything

They seemed to live on remembered things, to be so related to the seashore and the rocky hills and the loneliness that they are these things. To ask about the country is like asking about themselves. "How many toes have you?" "What, toes? Let's see – of course, ten. I have known them all my life, I never thought to count them. Of course it will rain to-night, I don't know why. Something in me tells me I will rain tonight. Of course, I am the whole thing, now that I think about it. I ought to know when I will rain."

—JOHN STEINBECK,
The Log from the Sea of Cortez, 63

"YOU GO WHERE IT hasn't been tilled, and you just don't get as many," Ed the boat driver says as he speeds us to Waiatt Bay, at the northern end of Quadra Island.

Anne Salomon, an experimental ecologist at Simon Fraser University (SFU), nods and glances at a chart on her iPhone.

"If you work the beach all the time," Ed says, "you get more clams."

Clams? After spending so much time in salmon central, I'm out with a team of ecologists and archaeologists, the scientific equivalent of the Rebel Alliance attempting to explode the Death Star. The Death Star, in this case, is the Pacific Coast's salmon-centric universe. The Rebel Alliance aims to reinvent the historical narrative. They dare to assert it took more than seasonal salmon runs to feed villages up and down the coast. The people cultivated the entire seascape and landscape. The world was much richer and more diverse than the salmon-centric worldview allows.

Oddly enough, one of the Rebel Alliance leaders has a name only one letter off from the name of the fish she plans to dethrone. "So, you think tilling the beach mimics clam habitat?" Salomon asks, her smile widening, a smile reflected in her green eyes. Salomon's skin is a duskier shade than most blondes and though it's May, she has the same base tan as other field biologists who spend much of their time outdoors. Not quite 40, the scientist could still be mistaken for a grad student.

"I would think so," Ed says. He has been a resident of Quadra Island since he was ten, but he's quick to point out he's still not a "local," even after 40 years or so. Like Salomon and other field biologists, the outdoors is painted on his face.

"You and I have the same hypothesis," says Salomon, as we skim through an area called Salmon Narrows. Waiatt Bay is accessible only by boat, or by foot from Kanish Bay, which is accessible from Granite Bay, where the road ends on the north side of the island. There used to be a fish farm in Salmon Narrows, directly in the path of migrating salmon, which is ironic, considering salmon already streamed by en masse.

"They moved the farm," Ed says, "but still, I don't think the blueback coho are coming back."

It's spring 2011 and the project that Salomon is launching at Waiatt Bay will test whether clam gardens – an ancient mariculture technique – were effective. Did they really nurture the growth of clams?

Archaeologist Dana Lepofsky roped Salomon into the project when the young scientist arrived at SFU in 2009 to establish her own lab. Lepofsky and geomorphologist John Harper, who had first identified the clam gardens as human-made in 1996 while on an aerial survey of the coast for the provincial government, showed some pictures to Salomon and explained their theory that these terraced beaches grew bigger clams than natural beaches. Salomon thought it was hogwash at first.

Harper and Lepofsky then took Salomon to Desolation Sound, Tla'amin territory only 42 kilometres away as the crow flies, southeast of where we stood on Quadra Island. "Dana is pointing out features here and there, I'm thinking, 'What's a feature?'" Salomon says. Her voice takes on a storyteller's cadence as she describes how she and her colleague Kristin Rowell, another experimental ecologist, thought the archaeologist was nuts. Harper caught Salomon and Rowell rolling their eyes behind Lepofsky's back. "Archaeologists are like geomorphologists," Harper told the pair. "They like to speculate wildly on very little data." But that speculation – thinking beyond cultural borders – can pay off.

Harper urged Salomon to go see the classic clam gardens on Quadra Island, and to hurry – she needed to view the beaches at the lowest tide. The next morning, clutching a map drawn by Harper, Salomon raced the tide. She hopped on a ferry, drove to a friend's parents' home to grab a canoe, loaded it onto the car's roof and shot off for Granite Bay with her friend's mother, 65, sitting shotgun. They arrived just as the tide changed. They threw the canoe in the water and paddled to Kanish Bay. "We were paddling, and I was saying 'come on, come on, come on' to myself," she says. "It took 20 minutes and then we saw at least six perfectly obvious clam gardens and my jaw dropped. It was an experimental biologist's dream come true."

Salomon and Rowell, an ecologist at the University of Washington, designed this experiment in 2011 – the first of its kind – that would tell them if the clam gardens did what the First Nations already knew they did through long experience.

I never imagined that hauling concrete cinder blocks onto a boat, and from a boat to a rocky outcrop, and from a rocky outcrop to a beachside camp, and from a camp to a canoe and from a canoe to another beach would amount to the lofty scientific life. But my sense of science is changing. Standing in cold water quantifying algae on rocks or the tree canopy above; performing open-heart surgery on fish; hauling garbage bags full of dead salmon around grizzly habitat – the quest for the puzzle pieces that might fit together to save the world can be quite athletic, laborious and chilly in this part of the world.

It's overcast but thankfully not raining when we get to Waiatt Bay. Ed drops us off, along with two canoes, four tents, six Rubbermaid containers and the dozen or so cinder blocks.

Graduate student Amy Groesbeck, undergraduate Brooke Davis, Salomon and I lug backpacks, coolers, tents, lifejackets, rope, tarps, Rubbermaid containers, sieves and cinder blocks to a clearing next to a creek. Water burbles past a wrecked trailer and a shack struggling to remain upright while a tangle of vegetation tugs it earthward. We lower the canoes into the water and paddle a few metres to the beach. We load the canoes with a couple of buckets full of rebar, spades, zip ties, Sharpies, clams, sieves – and those cinder blocks. No one complains, we joke that the blocks get heavier with each move. If we hurry – racing the tide, again – we'll have just enough time to visit a clam garden across the bay and plant some native little neck clams.

Salomon explains why the clams like being gardened. By creating a terrace, akin to a rice paddy, that floods as the tide comes in, the people who lived here extended the shore at the tidal height where clams thrive the most – the sweet spot where they grow fastest and live the longest.

The different Indigenous coastal cultures sustainably cultivated clams in the intertidal zone at least 1,700 years ago and probably longer. At the end of winter, while First Nations waited for the herring, followed by salmon and other foods, they could, even though low tides would be at night in winter, stroll down to the beach for fresh clams. Sleep is a more fluid thing in societies where the clock is tuned to natural rhythms so it's not a far-fetched scenario. Then the settler population changed traditional fisheries practices by outlawing fish traps, opening up clam beds to everyone, commodifying the catch and ending an ancient economic system of nonmonetary exchange.

In October 2010, I attended a Society of Environmentalists annual general meeting in Missoula,

Montana. At the meeting I asked Rebecca Miles, executive director of the Nez Perce tribe in Idaho, if part of the modern problem with salmon was our slavish devotion to the species, our elevation of its significance above all others?

The Nez Perce, an interior Salish tribe, has worked for decades to restore one of the most battered salmon runs in the Pacific Northwest, the chinook run on the Snake River. Before taking on the executive director position, Miles chaired the Columbia River Inter-Tribal Fish Commission. She knows a lot about fish and recovery efforts. Miles didn't seem surprised at the question and took a moment before she gave a thoughtful answer.

In the Pacific Northwest, First Peoples traditionally had a number of First Food ceremonies, depending on the environment, for deer, duck, berries and – significantly – salmon. Keeping a community fed was a complex, social enterprise, she said. Joe Hovenkotter, an environmental lawyer and staff attorney for the Confederated Salish and Kootenai tribes, who was listening to the question, interjected that designating an animal as a "keystone species" – the hub of a wheel to which other species are attached – is an umbrella strategy, designed to maximize protection of a habitat.

Does it really work? That status fails to acknowledge the fish as a part of the human community. As an umbrella it might hover over a few other species but it's not tied to the ecosystem, it has no threads. Nor does it take into account that humans must surely have been – and must surely remain – a keystone species, as intertwined with the ecosystem as salmon. Dee Cullon, Lepofsky and other archaeologists working in British Columbia struggle from year to year to map the material remains of the human imprint on the landscape, and their charts may still be sketchy. But a hardy, ancient fishery undoubtedly stretched along these shores. The archaeological record reveals human cultures evolving in tandem with salmon.

One of the older sites in BC, Glenrose Cannery, on the Fraser River, dates to more than 8,000 years ago.[66] The finds at Glenrose suggest that small groups of people roved across the landscape, seasonal locavores in search of the next good meal. When occupied, the site of Glenrose was on the oceanfront, but since 5,000 years ago, the shoreline shifted laterally and the site is now 20 kilometres from the sea. Artifacts – stone tools, sinker stones – slowly wash away into the current and the site is littered with fire-cracked rock. Archaeologists first excavated Glenrose in the 1970s, finding a menu heavy on salmon; eulachon; other small fish and shellfish, such as bay mussels; and deer and elk.[67]

At some point – the date is highly debated – the people of the northwestern coast of the Pacific Ocean developed an option: eat it fresh and move on or store it and stay. They chose to stay. In southeast Alaska, people stored food at least 4,000 years ago; on British Columbia's Central Coast, at least 7,000 years ago.[68] On the lower Fraser River, the change is noted about 5,500 years ago, roughly when sea levels settled down closer to what they are today.[69]

The population grew. Communities built permanent housing and developed trade networks and social hierarchies. More distinct cultures emerged. The Pacific Northwest coastal world came to look like what the Russian, English and Spanish explorers would encounter in the late 18th century. Complex communities occupied an ecologically diverse landscape. Six major language groups developed from north to south: Tlingit-Eyak-Athabaskan (to which the Haida belong); Tsimshianic; Wakashan (to which the Heiltsuk and Nuu-chah-nulth belong); Salish/Salishan (to which the Tla'amin and Sts'ailes belong); Chimakuan and

Penutian. All the coastal peoples retained strong ties to the marine world; salmon runs were important to all. Yet each group responded differently to the challenges of unique environments.

Amid the stormy, mercurial weather on the exposed archipelago of Haida Gwaii, people grappled with large expanses of rough ocean. Expert navigators, hearty fishers, the Haida were also daring. They paddled large cedar war canoes along the coast, raiding other tribes along the way, reaching as far as Washington state's Puget Sound to snatch Coast Salish people as slaves. In contrast, the protected inlets of Vancouver Island's west coast offered salmon and shellfish to the Nuu-chah-nuth, while the open waters brought them whales.

The Club Med of the coastal environment is where our little boat is taking us – where Salomon and the graduate students are seeking to unravel the secrets of the clam beds. It is the traditional territory of the Coast Salish, a community that dates, at least linguistically, to 3,000 years ago.[70] The Coast Salish bore the brunt of the European onslaught – disease decimated their communities, depopulating swathes of the landscape. Their environment also had the mildest climate and the most accessible resources, the 19th-century equivalent of an all-inclusive resort for families fleeing the daily grind back home.

Where early Europeans saw uncultivated "wilderness," the Coast Salish saw semicultivated patches of bulbs, berries and shoots – family-owned. Where Europeans saw rocky beaches miraculously stuffed by Mother Nature with clams, First Nations saw family-owned fish traps and clam gardens. Where Europeans saw red gold – infinite salmon runs that might earn them cash – First Nations saw a diverse fishery, family-owned, with carefully regulated distribution, the wealth continually re-assessed and allotted.

Socially designed resource management is not an anomaly. Risky places to live – thirsty deserts, tumultuous coastlines, polar landscapes – seem to beget caution on the part of human cultures. But until recently, only scientists studying hunter-gatherer societies were looking for communalism in the world's ancient cultures. On Fijian islands like Kadavu, until two generations ago, fishers boated to local coral reefs to feed their families. But by the 1970s, the fishing trips were taking longer and longer because the reef fish were becoming harder to find. Villages had abandoned an old system of no-fishing zones called *tabu*. Captain Cook heard the word when visiting Tonga in the 1770s, and it has been a part of the English language ever since: "taboo."[71] Such areas provided a reliable refuge for fish, a chance for populations to increase in size and number.

In the late 1990s, a Fijian marine biologist from Kadavu convinced the community to try a couple of *tabu* areas. Within five years, fish populations were increasing, along with predators like sharks and humphead wrasse (one of the largest coral reef fishes). This success convinced other villages to create community-controlled no-fishing zones; today they number more than 20.[72]

Before European contact, Hawaiian communities likewise had a complex set of rules that saw reefs regularly closed to fishing. There were strict regulations on fishing gear and bans on eating vulnerable species. The rules were based on something as unfair as gender and age, but, environmentally, it worked. The Hawaiians also built extensive fish ponds and stocked them with juvenile mullet and milkfish as insurance against famine. The Hawaiian Islands are at the mercy of tsunamis, floods, hurricanes and droughts, so their inhabitants required sophisticated risk management strategies. Some scientific interpretation of archaeological data puts the Hawaiian population at hundreds of thousands before European contact – there were definitely many thousands – and if it's higher, then the strategy worked spectacularly.[73]

In the Pacific Northwest, depending on salmon has always been a risky venture. Runs come and go, some failing for obvious reasons – the eruption of Mount St. Helen's, for example, which wiped out the salmon runs of many streams. But runs also fail for mysterious reasons. Why, for example, in the year 100 BCE, did sockeye disappear from a nursery lake on Kodiak Island in Alaska and stay away until 800 CE? Scientists who recorded the failure by analyzing lake sediments could find no catastrophic event to implicate.[74] Yet for 40 human generations, that run was gone. Alaskan Aboriginals were likely pondering how to overcome the loss when Jesus of Nazareth revolutionized religion in the Middle East. And then, when Charlemagne was crowned first Holy Roman Emperor in Rome, they were back.

Place yourself on the Pacific Coast in ancient times: If even the strongest of salmon runs are precarious, it's in your best interest to develop multiple food sources, including multiple salmon runs.

There are no ghost streams, no creeks too remote to bother fishing. And no salmon run is too small. Every beach, every watershed, every run is important. Some territories are richer than others, of course. When the wealth is salmon, a food that sticks to the ribs and preserves well, it's prudent to make friends with the neighbours. Or raid them. Or push them out.

This was no Eden, no innocent land before the advent of evil. It was a chaotic environment with an insecure food supply. Earthquakes rumbled, volcanoes burst, fish runs failed, diseases killed. There were social inequalities, there were wars, slaves were taken. People adapted to fat times and lean times, to war and peace. When the Europeans arrived in the 18th century, they found bands of the Laich-Kwil-Tach warring with the K'ómoks (Comox) to the south. The K'ómoks eventually retreated, giving up territory, and the groups formalized the end of hostilities with a marriage.[75]

Intermarriage was also one of the ways the Coast Salish and other tribes gained access to richer salmon runs. People from Vancouver Island with ties to Fraser River clans travelled regularly in the summer to fish the Fraser and its tributaries. Complex social and fishing rules regulated their interactions. The geographically fortunate Sts'ailes (formerly Chehalis Indian Band) did not have to leave their territory around the Harrison River to find food among related clans, but they had to enforce a code, a selective fishing model, on themselves, their neighbours and their relatives. Food, after all, was wealth.

So here we are, the four of us – me, the writer; Salomon, the university professor; and Groesbeck and Davis, the two young students, each of us certain science can change the world for good – sitting on a beach, the remnants of wealth invested before us, a clam garden. A simple rock barrier and a small cleared beach. We sort the almost 200 clams before us, stacked in neat piles of similar sizes like cookies in a display case, except this display case is a slab of rock. Salomon weighs the labelled clams and I record numbers in a notebook. The clouds scurry off, replaced by sun.

The students – Davis's blond ponytail swings freely from the back of the ball cap on her head, Groesbeck's thick braid of red hair swings over one shoulder – jump into a green canoe to plant clams at a garden nearby, a five-minute paddle around a bend. Salomon and I tackle the one before us, a small beach, its rocky barrier partly dismantled.

We carry over six rigid mesh bags, 15 clams in each. We drag over six pieces of rebar, a cinder block, rope, flagging tape and spades for digging. We plant one mesh bag subtidally, tied to the cinder block to keep it in place. The others we space up the beach at different points in the intertidal zone. We bury each bag flat, a handbreadth below the surface, and zip tie it to the rebar we pound into the beach with a rock. Rebar and cinder blocks aside, it's a lot like gardening. We're just finishing when Davis and Groesbeck paddle back. Two beaches down, nine to go.

We set a comfortable pace back to camp. The flat water seems to part, making way for the canoe. We lift our paddles as the boats nose onto the beach, a soft, crunchy noise telling us we can jump out and pull the canoes up to dry land. It's still light out, so we hurry to set up tents and sleeping bags and a camp kitchen. Pink and orange bands of light announce the sun's departure, prompting Salomon and Davis to start dinner, cooking curried vegetables and rice as Groesbeck and I haul a white plastic bucket full of live clams out of the water. We start sorting. After eating and chatting – about careers, balancing work and family, work worth doing – we all sort and label more clams. With a woodworker's metal file, we create rough patches on each shell for tags to stick to, then weigh the bivalves on the small digital scale you might find in a foodie's kitchen. Clams lack brains but they do know when something's not right, when a predator like a starfish is after them. They use a muscular foot to burrow into the sand away from predators. What do they make of this rasping across their shells?

The sky fills with stars. We pack away the gear and climb into our tents. The next day we rise with the sun and do it all over again. Except it's raining. For a couple of hours, Salomon and I huddle in her tent, while the two students huddle in Groesbeck's. We dry off clams with our wool socks to make sure the labels can stick and place them on a black wool sweater; they look like white and milk chocolates nestled in a box. We quickly deplete a roll of orange tags. Salomon saved the tags from studying reefs in New Zealand a few years ago; at $150 a roll, they were too expensive to leave behind and now, they are too expensive to risk mangling.

"You know," Salomon says, carefully perching a clam on the scale. She looks at me, her eyes go wide and she smiles. "Here we are a group of women, processing clams. If we were here a few hundred years ago, we'd be doing the same thing," she says.

Each lab works something like a fish camp – I know that – yet when Salomon says it, I grab the notebook beside me and start scribbling as I nod in agreement. But it's more than a bunch of women processing clams. It's a scientific lab as an apparition of the past, a reincarnation of traditional ecological knowledge and research. It's our modern way of mimicking an era when humans had to read their local environment to live. These scientists are neo-Indigenous: they attempt to live this world, to think beyond their own experiences.

Reading the environment means starting with a vocabulary and – yes – categories. Carolus Linneus, the 18th-century taxonomist, revolutionized biological classification, but it's not something humans invented. Anthropological research suggests that all societies created a vocabulary and categories, that taxonomy is hard-wired.[76] We organize organisms in similar ways across cultures. We use inductive reasoning – applying categories – when we come upon an unfamiliar habitat. We need cognitive tools to deal with novel places.

Yet practical knowledge of nature is all but gone in the general population of educated, industrialized societies. The cognitive map of ecological relationships – relationships that include humans – is all but erased. Few of us learn that the clam I have in my hands is of the kingdom Animalia, the phylum Mollusca, the class Bivalvia, the order Veneroida, the family Veneridae. We leave that to the kind of scientists I'm following in the field, or to what some anthropologists call "folk" biologists and others call "knowledge holders." These latter are the Sts'ailes fishers who told me when the different salmon species run and birds gather, the ex-fishers aboard the *Ricker* who identified unfamiliar fish, Carol Schmitt who has played with chinook salmon for decades and Ed, the skipper who explained to Salomon what he looks for in a clam bed. The ecologist sitting across from me in the tent, one hand holding a clam, the other hand tucked into a wool sock she skims over the shell, can read this environment better than me. Salomon grew up on the BC coast, sailing, learning how to dive and allowing her curious mind to lead her on all sorts of watery adventures. But more importantly, she spent years in formal educational settings in different parts of the world. She got better and better, as her education progressed, in applying what she was learning to the natural world.

Like the other scientists fanning out across the BC coast, Salomon also gathers knowledge from folk biologists.

Anthropological research has drawn parallels between Indigenous Elders and biological "experts" in North America, both are groups of people with at least 20 years of experience in their fields.[77] Place a North American, formally educated ornithologist – or an experienced amateur birder – in Itza' Maya territory in Guatemala and he or she can order the rainforest in similar fashion to the Itza' Maya who depend upon it. From studies, it can also be assumed that if an Itza' man or woman came to Stanley Park in Vancouver, he or she could order the natural world in similar fashion to the North American expert. That is to say, the Itza' Maya would do far better than the average Vancouverite, whose grasp of either environment – rainforest or city park – is rudimentary.

Within the rainforest, the Itza' Maya also have a better grasp of ecological relationships – explaining relationships as reciprocal between plants and animals – than the two other communities with whom they share territory: Ladinos, who are Spanish-speaking and of mixed Indigenous and European descent; and Q'eqchi', Maya who immigrated from the highlands, herded by Spanish colonialists a couple hundred years ago. The Itza' Maya, in contrast, have lived in their world for generations, culturally evolving with the landscape.[78]

Without scientific labs and folk biologists – whether Indigenous or settler – we would be doomed right now. And for many of us, the only way to gain an ecological education quickly is through an expensive university program.

The four of us – Salomon, Groesbeck, Davis and I – spend the afternoon canoeing from cove to cove, seeding three more gardens. At my favourite, a rock wall as long as two tennis courts rises chest-high, straight out of the water, forming a long, flat, cleared beach: perfect clam habitat.

A telltale sign of good clam habitat is shell hash, layers of broken shell. It's not just ancient trash. The old shell sends chemical messages to juvenile clams, saying something like, "Live here, plenty of food!" This is a habitat where they'll find lots of plankton to siphon up, providing nutrients to grow big. At least, that's what scientists who study oysters have conjectured about that species,[79] and clam scientists like Salomon are intrigued.

During two days of "gardening," we plant 450 juvenile clams at five beaches. We would plant more, but it's time to meet up with the archaeologists surveying Kanish Bay. This project is interdisciplinary, archaeology and ecology working in tandem to puzzle out the past. We pack, either leaving the cinder blocks resting in the subtidal zone holding down some of the clam cages or stashing them away, tucked among the trees, nestled on mosses amongst the ferns, ready for the next planting session. Groesbeck climbs into a canoe to retrieve from the bay the live clams she has yet to tag, weigh and record. A yellow rope, tied to rock on shore, snakes into the water. She paddles to where it disappears, peering into the water. She turns to look over the other side, then paddles a little more. The bucket should be at the end of the yellow rope. It's not.

I know one knot, and I used it to tie the rope to the bucket handle. Maybe it wasn't the appropriate knot. Or maybe I forgot how to tie a bowline knot: if I can't remember how the rabbit is supposed to go into its hole, the knot will only be right one out of two times. I've lost the clams.

Salomon sheds her clothes and dives into the chilly spring waters. It's too murky. She can't see the yellow rope. She swims in circles, dives, comes up, repeats a few times, gives up, and wades ashore, shivering, clamless. The least I can do is donate my dry, clean towel.

A hummingbird, its throat a ruby red, buzzes the archaeologists and ecologists, ten of them sprawled in front of me on a beach of seashells and clams. Their leader, Dana Lepofsky, the archaeologist who first recruited Salomon to study clams, looks up at the sound of whirring wings. "Come see me," she says, holding out her arms to show off her purple jacket and matching nail polish. "Look what I'm wearing." Lepofsky, a slender, fit, 50-something-year-old, has thick, shoulder-length, dark hair, her age only reflected by a few strands of silver. Her voice is soft and sweet, nurturing.

The hummingbird levitates up and away, a sign of summer teasing the researchers, who have stopped chatting about herring, anchovies and clams to admire it. The past week of fieldwork has been mostly rainy and cold; a hummingbird seems a messenger from another world. But sunshine is peeking through the clouds, illuminating a thick layer of white shell bits in the sand. Maybe the bird has indeed brought summer.

The beach where we're sprawled is really a midden on a small island – nicknamed Shell Island by this team – in Kanish Bay. When Ed dropped us back at the wharf, we loaded our cars with the canoes and other gear and drove to Granite Bay. We unpacked, hauled gear across a muddy beach at low tide, packed the boats and canoed to Shell Island in time for a short paddle to see some clam gardens. Then it's dinner, a campfire and marshmallows.

On the beach, ecologists and archaeologists, our red, orange, blue and brown tents crowning the hill above us, discuss the project. Salomon leans back on her elbows, legs stretched out, as Lepofsky winds up a mini-lecture: "When a relatively nontypical species pops up in the archaeological record," Salomon asks, "how do we know it's not traded? Let's say anchovy." The graduate students cock their heads, pondering; nobody ventures an answer.

"Now it's a bit funny, I think, to ask the question in the first place," Lepofsky says, sifting shells idly with a tanned hand. "People don't ask that question about salmon." She pauses. "Well, this is me being defensive," she admits. "Definitely a lot of trade happened." Like all good lecturers, Lepofsky creates a monologue, complete with imaginary challengers to prod the weak points in her reasoning. The point she is spooling out this morning is that the little guys, herring, eulachon, anchovies and clams – the species that are neglected on the coast right now – went a long way toward feeding ancient villages. They were traded, but they were also, overwhelmingly, fished locally.

Because of its sheer abundance, salmon tends to overwhelm the archaeological record. But evidence is piling up that the little fishes – along with fish roe, waterfowl and clams – went a long way toward feeding entire villages.[80] The ancient maritime economy was a managed and diverse biological buffet. Salmon was important, but, as Lepofsky argues while we loll on the beach, its proportion on the plate has probably been inflated. The picture emerging from her work – a collaboration with many scientists and First Nations – is one of a culture that avoided relying on a single, plentiful species. It's almost as if people hedged their bets when it came to food; a risky environment demands caution.

Salmon is very much present in the record, Lepofsky says, but it doesn't appear to have gone through a classic boom-and-bust cycle of overexploitation. The prehistoric cultures had to be conservationist, and their cautious attitudes survived the European invasion – even if those attitudes didn't triumph in the European-dominated culture that emerged. "If you talk to the Elders about these intertidal sites," Lepofsky says, "they don't really think about them as single-purpose the way we tend to."

In the 1930s, anthropologist Homer Barnett spent time with the Tla'amin and wrote about their fishery – fish traps, nets, hooks and harpoons for salmon, halibut and cod – and the prized delicacy of sea mammals. The clams, he wrote, "were everywhere and a never failing source of food."[81] Barnett also witnessed the herring harvest: two people would go out in a canoe, often a husband and wife. The woman paddled and the man used a herring rake to impale the fish. They collected herring by the hundreds or more. For the roe, the Tla'amin submerged branches of Douglas fir, hemlock or red cedar near the shore. The herring deposited their eggs and the people gathered them. Tee-sho-shum, one of the Tla'amin's main villages, translates to "water white with herring roe." Barnett notes, offhandedly, that Tla'amin and two other Salish tribes had no first salmon ceremony.[82]

Lepofsky is a big Barnett fan. She hunts down copies of "The Nature and Function of the Potlatch," his 1938 PhD dissertation from the University of

Oregon, as gifts for her graduating students. The manuscript outlines an economic system almost completely at odds with the industrialized, capitalist system brought by the European settlers.

Archaeologists studying BC's coast will spend the winter screening the samples they take from beaches like this one, to which we've come to investigate clamming. Through the tedium of screening, they will figure out the size of the fish – and the species – taken by ancient communities. It takes hours to sift dirt through a screen made of two-millimetre mesh, picking out herring and other fish bones from the debris and sorting the remains of clam shells, cockles and sea urchins by the hundreds. Show a tiny fish bone to toiling archaeology graduate students and they will usually identify it in a flash – "cleithrum, probably from a herring, part of the pectoral girdle, it sits behind the gills." They'll eventually write the kind of theses that could help to dethrone salmon, at least among scholars, finding, say, that herring overwhelms salmon in the faunal record from certain beach samples.

On some beaches researchers are finding chevron- and heart-shaped traps from more than 1,000 years ago, traps that massively harvested small fish by funnelling them into woven corrals. On others they are identifying species no longer found in the ecosystem, such as native oysters. They are noticing that when the remains of one fish decreases – salmon, for instance – the remains of another increases. This is the kind of evidence that helps show the character of ancient humans. Adapting to a constantly shifting ecosystem, they valued both large and small fish.

In the past, archaeologists used screens with no smaller than six-millimetre mesh. But that is way too big to catch the bones of herring, anchovy and eulachon. Mesh even half that size will miss the smallest remains. So archaeologists like Lepofsky – and others working in Alaska, Haida Gwaii and Oregon – have what amounts to an unspoken agreement to redress a myopic fixation on salmon, through mesh size. Using nested screens as small as two millimetres, they compare the fish bones they recover. Slowly, bone by bone, they are changing the record, finding the smaller fish, the fish that have been wiped out, the fish whose presence was once missed in the tangle of invasive species of recent years.

The downside of the commitment to small mesh is that it quadruples the screening time and costs; funding availability, as always, skews science. But fighting for funds is part of the job. Or, as an archaeologist might put it, fierce competition for dollars is the way the current dominant culture works. Scientific labs are no exception.

The hummingbird was indeed a good sign. The sun shines brighter and the water stays calm. The ecologists leave to dig more clams for Groesbeck's study; the archaeologists drag canoes and kayaks to the water for a reconnaissance mission. I climb into a canoe with Louie Wilson, who I first met in Campbell River, when he worked with Dee Cullon a year earlier.

Lepofsky's target is two types of sites: shallow and dispersed layers of clamshells, indicating places where people shucked clams and brought the meat back home; and complex settlement sites that likely predate the clam gardens. What she hopes to find is a transition in the middens between pre- and post-clam garden. She'll note the abundance, age and size of various clam species harvested in various eras and get an idea of when clam harvesting became more managed and whether innovations in ancient technologies increased the size and number of clams. This is where the ecologists' work dovetails with the archaeologists.

We paddle into a deep cove. As we pull our canoes up onto the beach, an osprey wheels and hovers overhead. We've paddled maybe 20 minutes from Shell Island.

"If there's a transition, if there is some change in productivity, hopefully that will mirror the results of the hypotheses being generated by the ecology," Lepofsky says as we head toward the trees. "If that's mirrored in the midden, then we'd be able to date that transition."

Ahead of us, a few graduate students troop up the beach and are swallowed by cedar and fir trees. We follow. We're looking for a settlement midden to sample. It should be a breeze. In coastal BC forests, shells often speckle the bottoms of uprooted trees and cling to creek faces, sometimes far upstream from the shore. I think we're meandering, but Lepofsky looks for a feature, something that tells her people had a hand in shaping this environment. Before long, she finds a ledge. She rummages two trowels from a plastic box and hands one to Megan Caldwell, a graduate student. Caldwell has been at this a while. Her master's thesis focused on the fish traps in Comox Harbour, about 50 kilometres south. Before excavating a 2 x 2 cm column sample, 70 cm deep, Caldwell runs through the routine with Michelle Puckett, a new grad student. Caldwell is quick, she needs to catch a ferry for another site across the Georgia Strait, in Tla'amin territory, by early morning, when the tide is at its lowest.

"I might not have anything else in my life organized, but I know when the tides are," she jokes.

Notebook in hand, Puckett draws the layers revealed by slicing away the soil. We also take photos. Puckett collects soil samples in plastic bags, passing them to Lepofsky, who sits above the ledge. Lepofsky, who has a PhD from the University of California, Berkeley, and has worked in the storied American Southwest and Pacific Islands, tells me about her first brush with the Pacific Northwest, in the Central Coast studying the traditional use of plants. She loved it and never left.

We spend the next few days in the bush looking for the past, gathering at night for dinner and storytelling around the fire. One day most of us have to get back to daily life and we leave Groesbeck and a few others behind, leaving the extra food we brought, water and some gear if it's needed.

"Potlatch" is an Anglicized version of the Nuu-chah-nulth word *pachitle*, a ceremonial feast with two huge benefits for the people who practised it. Potlatch ceremonies both redistributed wealth and maintained a healthy ecosystem and are in part what made the Pacific Northwest Coast famous in anthropology circles. The more a community had to give away during potlatch ceremonies, the richer the community. The wealthiest had access and rights to exploit natural resources, but they also had the responsibility of regulating the take to ensure a steady supply to the community. Potlatching was practised up and down the coast, each community had its own variations, but the core ideal – reciprocity – was the same.

After spending almost a week on Quadra with Salomon and Lepofsky, I had to read Barnett's dissertation. "The Nature and Function of the Potlatch" is not painful to read. If Barnett is any indication, researching and writing a dissertation was more fun in 1938 than it is today. No slogging through a long genealogy of anthropological writing, redefining every anthropological term ever found in print. Barnett simply dips the readers into a world, in this case Coast Salish, where wealth came from the land and sea, the rich were expected to "pay" what they could afford and the people expected everyone to give back as much or more than was received. In the reciprocal economic system of the world Barnett found, generosity was highly rewarded. "Obviously, the system works only because everybody – everybody who is included in it – has the same idea; namely to give back as much or more than has been received," Barnett wrote. But this reciprocity, oddly, occasionally begot rivalry, too, with rival potlatches becoming contests of destruction. And for people with slave origins, no matter how many potlatches they hosted, they would never rise above their ancestry. No matter the culture, each of us can fall victim to pride. And we also can't resist play, something else Barnett found in the potlatch, with wrestling, racing and eating contests, as well as clowning around and comedy. "Sometimes the food intended for one man is eaten by those who pass it along," Barnett writes. And even a potlatch for the dead is "interspersed with rollicksome incidents."[83]

It sounds a lot like a scientific lab.

I reread Barnett as I was getting ready to visit Megan Caldwell at one of her Tla'amin sites in the same spring, in 2011, and he recharged my sense of wonder at the grand archaeological project.

I catch up with Caldwell at Gibsons Beach, a beach Barnett likely strolled as well. Barnett interviewed Tla'amin Elders, none of them are alive today, but their grandchildren are and I'm here to chat with them. Caldwell, Nyra Chalmer, another graduate student, and I measure and map some traps. These are stone traps, different from the traps in Heiltsuk territory. And this is what puzzled Salomon when she saw her first clam gardens in Desolation Sound, not far from Gibsons Beach. The Tla'amin may have had dual-purpose technology – traps and clam gardens intertwined.

In Desolation Sound, Salomon saw what looked like a jumble of fish traps and terraces for the herring and salmon runs, along with the clam resource. Waiatt Bay has no fish run. It was perfect for gardening clams, and the people likely honed the technology to that environment. Likewise the environment here dictated the technology.

Survey done, we pack up and drive five minutes to the Tla'amin main village. We stroll the front "yard," a jumble of sunken fish traps that echo the pre-Colombian-era remains we've just measured at Gibsons Beach but are famous among archaeologists because they're so extensive and obvious. Stone walls that form traps scallop the shore for a kilometre – these traps still catch salmon and other species in season. A breeze riffles the water, disrupting the sun's warmth, chilling our bare arms. Twin markers of the relentless march of the outside world frame the site: a church squats on the waterfront and, in the distance, a paper mill bellows steam.

We walk up the beach and onto the road to the band office where Michelle Washington, the land use planning coordinator for the Sliammon Treaty Society, has worked with the SFU archaeologists from the beginning of excavations in the territory. She introduces herself as Washington but also by her traditional name, Siemthlut. She introduces me to her co-workers and gathers up some coffee mugs for us as we retreat to another room. Washington's brown eyes sparkle. She has an open face, smiles easily and has a husky laugh that's a gift to those around her. Her long, black hair is swept up in a ponytail that streams down her back.

Washington is only 40, young enough that in many families she might have grown up ignorant of Tla'amin traditions, but as a girl, she was curious about the old ways and spent time with her grandparents. Over the years, she interviewed lots of Elders, learning about the interplay of the environment and the people: stone traps invited in the herring precisely when salmon stores were low (early spring) and villagers needed an influx of fresh food.

"When I was a kid, this time of year was alive, this village was alive," she says, placing a couple of thick manuscripts she grabs from a shelf on the table. One is a land use survey she authored. "Everybody had their smokehouse ready, their herring racks ready, the kids were all down the beach, you know little kids just had ice cream buckets, the bigger kids had bigger buckets, because you could go down into the little pond in the intertidal zone and scoop up hundreds of herring in one scoop and run it to your house. That's not that long ago: 1984 was the last year I remember that happening."

But the damage to Indigenous fisheries began long before that. And when people stop gathering or cultivating a species – like herring – information is lost forever. Especially when language is also lost. Anthropological writings, like Barnett's, fill in gaps and confirm oral histories – though most early researchers failed to see signs of active management. Washington is keen to make sure the contemporary story of her people becomes part of the historical narrative of the coast. The Tla'amin weren't just lucky, endowed with mild weather and abundant resources, she says, they worked the land according to ecological principles developed by their ancestors. They cultivated careful relationships, both nonhuman and human.

High-class Coast Salish families inherited rights to abundant salmon runs, and they consolidated wealth by marrying other elites. *Heh goos* – "head men" in the Tla'amin language – made decisions on when and how to fish, and their status was legitimized through the public ceremonies of the potlatch, when they gave away their wealth. The lower-class Coast Salish had little social mobility. Status, for the most part, was inherited and changes in the social hierarchy were rare. The lower class had a lot of incentive to cooperate with the wealthy. Soon after potlatches were outlawed in 1885, the chief of the Kwakwaka'wakw – ancestors of the tribes belonging to the Laich-Kwil-Tach Treaty Society – told anthropologist Franz Boas: "It is a strict law that bids us distribute our property among our friends and neighbours. It is a good law."[84]

Potlatches not only distributed wealth – a finite natural resource – they also distributed knowledge, which is not finite unless it gets lost. That's always a danger, especially if it's not written down. Washington is teary when she talks about knowledge disappearing. The Coast Salish way of living was hard won and will not easily be retrieved. "My Granny used to say something that I never quite understood until I got older," Washington says. "She would look at expensive homes with manicured lawns and say, in our language, 'Oh those poor people, they have no medicines or food in their yard. How are they going to feed themselves and take care of themselves if anything happens?'"

The scientific work on sockeye, chinook, coho, chum and pink, and on their relationship with their environments, is painstaking, expensive and highly structured. It is also the only means of translating Indigenous and community knowledge into the language of the dominant culture. Science gives language to a way of living that requires multiple means of communication but often few numbers or words. For the Coast Salish Elders, to know salmon – or any other fish, shrub or tree – was to think like them, and to think like them required knowing them – a circular path. What's required is daily observation. The act of urbanizing removes us from intimate contact with the natural world, interrupting the circle.

Seceding from that kind of knowledge is dangerous.

The dominion government banned the potlatch to open the way for a new, capitalist system of rewards and punishments, and yet scientific labs today abide by their own kind of potlatch. In the corner of our culture that addresses the urgent problems of disease, climate change and food scarcity, we still value sharing. Successful labs have generous principal investigators – our version of *Heh goos* – who have access to knowledge, equipment, funds and other "elite" principal investigators in the same field. Generous principal investigators beget generous graduate students, who move on to establish their own labs, or work in industry or in government, and keep ties with their old labs (that's the intermarriage part). A lab's generosity and reciprocity make it resilient. As long as society values the knowledge a lab produces, it remains prestigious – and stable. This is sometimes why scientific theories that need to die might take a long time to do so – they come from a resilient lab. In such cases, a theory only dies when the investigators die. And anyone who has spent time in academia also knows it's full of the best about humans (generosity, cooperation and fun), as well as the worst, often rife with unhealthy rivalries and too much pride.

Scientists dislike hearing that what they do is a cultural endeavour; they think it's objective, non-biased and above human emotion. But if we only considered the cardiovascular data from the sock-eye corralled by Scott Hinch's lab, the DNA from the juvenile salmon caught in Marc Trudel's net, the number and size of the stakes Dee Cullon records, the number of pink and chum salmon that reappear in John Reynolds's study streams, how big clams might grow in a clam garden Anne Salomon tends or the varied fish bones Dana Lepofsky finds in a sieve, it is reality, as Steinbeck says, but maybe the least important reality concerning the salmon.

What strikes me as most important is how each lab reflects a way of being in the world, their economy of knowledge a reciprocal economic system that humans fall back on again and again when wealth is tied to something other than the abstract idea of money, freed from the marketplace. An imperfect system because humans are imperfect – and there will always be cheats like the chum satellite male – but a system with straightforward rules, sometimes as simple as "you get what you give."

The salmon have no narrative without humans chasing after them. We create their narrative and today it runs from bleak to bleaker, from rampant human induced viruses to extinction. But after my own chase, I just don't feel that alarmed. Maybe I should. Yet after six million years on Earth, after over 12,000 years circling the Pacific Ocean, this fish seems better equipped to handle the uncertainty of climate change than we do.

Climate change has shaped all of Earth's fauna, determining which species have gone extinct and which have survived. From the stout infantfish, the smallest known vertebrate, to the blue whale, the largest, we are all subject to the forces of natural selection, in which climate plays a huge role. I asked one population geneticist how he would fix declining sockeye runs and he said, "Probably just fix their habitats and leave them alone." Habitat, of course, includes climate. Will salmon survive? I'm pretty sure Charles Darwin would weigh in with: "I can see no limit to this power, in slowly and beautifully adapting each form to the most complex relations of life."[85] Or, as the famous paraphrase goes: "It is not the most intellectual of the species that survives; it is not the strongest that survives; but the species that survives is the one that is able best to adapt and adjust to the changing environment in which it finds itself."[86]

That's another reality straight scientific data

cannot tell us – considering the risky environment we're facing, do we have the knowledge we need to adapt?

Washington and I look over a map she has in her office. Pins mark traditional areas of fishing, hunting and foraging in Tla'amin territory, points on a landscape that contribute to a narrative, an interrupted narrative that is about more than salmon, a narrative that needs each of us to stitch it back together.

NOTES

1 Steinbeck, *The Log from the Sea of Cortez*, 1–3.

2 Ibid., 2.

3 Trosper, "Resilience, Reciprocity and Ecological Economics," 208.

4 Blount et al., "Neonatal Nutrition, Adult Antioxidant Defences and Sexual Attractiveness in Zebra Finch"; McGraw and Ardia, "Carotenoids, Immunocompetence, and the Information Content of Sexual Colors"; American Association for the Advancement of Science, "Bright Beaks Provide Honest Look at Male Immune System, *Science* Study Says."

5 Lackey, "Salmon Decline in Western North America"; Government of British Columbia, *2010 Year in Review*; Hinch and Martins, *A Review of Potential Climate Change Effects*.

6 Lewes, "Crisis in the Fur Trade," 244; Maclachlan and Suttles, *The Fort Langley Journals*, 285.

7 Hendry et al., "Rapid Evolution of Reproductive Isolation in the Wild."

8 Quinn, *The Behavior and Ecology of Pacific Salmon and Trout*, 47–48, 222–223; Quinn, Merrill, and Brannon, "Magnetic Field Detection in Sockeye Salmon."

9 Scheinfeldt et al., "Genetic Adaptation to High Altitude in the Ethiopian Highlands"; Beall, "Andean, Tibetan, and Ethiopian Patterns of Adaptation to High-Altitude Hypoxia"; Beall et al., "An Ethiopian Pattern of Human Adaptation to High-Altitude Hypoxia."

10 Quinn, *The Behavior and Ecology of Pacific Salmon and Trout*, 277; Gende et al., "Pacific Salmon in Aquatic and Terrestrial Ecosystems."

11 Walters and Maguire, "Lessons for Stock Assessment from the Northern Cod Collapse"; Rose, "Reconciling Overfishing and Climate Change with Stock Dynamics"; Hutchings, "Spatial and Temporal Variation in the Density of Northern Cod."

12 Dietz, Ostrom, and Stern, "The Struggle to Govern the Commons"; Waddy and Aiken, "Multiple Fertilization and Consecutive Spawning in Large American Lobsters"; Shannon Tibbetts (DFO research biologist), personal communication with author, May 14, 2014.

13 Martins et al., "Effects of River Temperature and Climate Warming on Stock-Specific Survival"; Hinch and Martins, *A Review of Potential Climate Change Effects*, 5.

14 Asche et al., "The Salmon Disease Crisis in Chile."

15 Miller et al., "Genomic Signatures Predict Migration and Spawning Failure in Wild Canadian Salmon."

16 Hinch and Martins, *A Review of Potential Climate Change Effects*, 4–5.

17 Kocan, Hershberger, and Winton, "Ichthyophoniasis"; Zuray, Kocan, and Hershberger, "Synchronous Cycling

of Ichthyophoniasis with Chinook Salmon Density"; Richard M. Kocan (professor emeritus at University of Washington), personal communication with author, August 8, 2011.

18 Richard M. Kocan (professor emeritus at University of Washington), personal communication with author, August 8, 2011; Kocan and Hershberger, "Differences in Ichthyophonus Prevalence and Infection Severity."

19 Schindler et al. "Population Diversity and the Portfolio Effect in an Exploited Species."

20 Mackie et al., "Early Environments and Archaeology of Coastal British Columbia."

21 Martin, "The Discovery of America"; Martin, "Prehistoric Overkill"; Grayson and Meltzer, "A Requiem for North American Overkill."

22 Hayashida, "Archaeology, Ecological History, and Conservation."

23 Van Duyn Southworth, *A First Book in American History with European Beginnings*.

24 Ripple Rock Explosion 1958; Wright et al., "Observations on the Explosion at Ripple Rock."

25 Connaway, *Fishweirs*, 24, 47, 313, 333, 366; Erickson, "The Domesticated Landscapes of the Bolivian Amazon"; Cooney, "Introduction"; English Heritage, *Introductions to Heritage Assets*.

26 John Steckley (Humber College), personal communication with author, December 4, 2013.

27 Maclachlan and Suttles, *The Fort Langley Journals*, 178; Parks Canada, "Fort Langley National Historic Site of Canada."

28 Parks Canada, "Fort Langley National Historic Site of Canada"; Omega Pacific, "Culturing Pacific Salmon since 1980"; MacKinlay et al., *Pacific Salmon Hatcheries in British Columbia*.

29 White/Xanius, "Heiltsuk Stone Fish Traps."

30 Hoffmann, "Economic Development and Aquatic Ecosystems in Medieval Europe"; Roberts, *The Unnatural History of the Sea*, 24.

31 Connaway, *Fishweirs*, 47–55.

32 Roberts, *The Unnatural History of the Sea*, 22–24, 220–221; Hoffmann, "Economic Development and Aquatic Ecosystems in Medieval Europe"; Hoffmann, "A Brief History of Aquatic Resource Use in Medieval Europe."

33 Connaway, *Fishweirs*, 47–55; National Archives, "Salmon and Freshwater Fisheries Act 1975"; "The Magna Carta Project"; British Library, "Treasures in Full Magna Carta."

34 Connaway, *Fishweirs*, 47–55.

35 Danziger and Gillingham, *1215*.

36 Boas, *Indian Myths & Legends*, 616–617.

37 Quinn, *The Behavior and Ecology of Pacific Salmon and Trout*, 13.

38 After talking to many elderly people over the years in British Columbia, I've learned that they grew up calling pink salmon "cat food," partly because pinks are cheap,

dirt cheap when fresh, and most commonly canned. See Encyclopedia of Life, "*Oncorhynchus gorbuscha*." The state of Alaska acknowledges that they are the least valuable commercially, which means they are turned into processed fish food (not necessarily for cats). See State of Alaska, "Pink Salmon (*Oncorhynchus gorbuscha*)."

39 Kentaro Morita (Hokkaido National Fisheries Research Institute), personal communication with author, June 8, 2010.

40 Karpenko and Koval, "Feeding Strategies and Trends in Pink and Chum Salmon"; Vladimir Radchenko, personal communication with author, June 2010; Hinch and Martins, *A Review of Potential Climate Change Effects*; Pacific Salmon Commission, *Report of the Fraser River Panel*.

41 Hocking, Ring, and Reimchen, "Burying Beetle Nicrophorus Investigator Reproduction"; Hocking and Reynolds, "Impacts of Salmon on Riparian Plant Diversity"; Hocking and Reynolds, "Nitrogen Uptake by Plants Subsidized by Pacific Salmon Carcasses."

42 Isabella, "Wrens May Have Salmon to Thank for Breeding Boost."

43 Huxley, "Symbiosis between Ants and Epiphytes."

44 Araki et al., "Fitness of Hatchery-Reared Salmonids in the Wild."

45 Elroy White, personal communication with author, Heiltsuk Territory, Central Coast, British Columbia, May 2012 and May 2014.

46 Steven Schroder, personal communication with author, April 1, 2014.

47 Ibid.

48 Berghe and Gross, "Female Size and Nest Depth in Coho Salmon (*Oncorhynchus kisutch*)"; Steven Schroder, personal communication with author, April 1, 2014.

49 Quinn, *The Behavior and Ecology of Pacific Salmon and Trout*, 73.

50 Reimchen, "Some Considerations in Salmon Management."

51 Charles Darwin, *The Life and Letters of Charles Darwin.*

52 Groot and Margolis, *Pacific Salmon Life Histories*, 233; Miyakoshi et al., "Historical and Current Hatchery Programs"; North Pacific Anadromous Fish Commission, "Rehabilitation of Japanese Salmon Hatcheries."

53 Knapp, "Trends in Pink and Chum Salmon Markets."

54 The Inner South Coast, West Coast Vancouver Island and Fraser River chum salmon fisheries received MSC certification following an independent, third-party, rigorous scientific assessment that was conducted by Intertek Moody Marine. A fourth unit of certification, the North/Central Coast fishery, is still in assessment. See Marine Stewardship Council, "Three British Columbia Chum Salmon Fisheries Awarded MSC Certification."

55 Census of Marine Life, "Niskin Bottles"; Encyclopædia Britannica, "Niskin Bottle"; NASA, "Spinoff Database Record. Spinoff from a Moon Tool."

56 Northwest Fisheries Science Center, "Copepod Biodiversity."

57 Knapp, "The Wild Salmon Industry."

58 Curtis, *The North American Indian*, 246–247.

59 Quinn, *The Behavior and Ecology of Pacific Salmon and Trout*, 149, 195.

60 Parenti, "On the Status of Species."

61 Lathan, "Rubber Gloves Redux."

62 Ereshefsky, "Darwin's Solution to the Species Problem"; Darwin, *On the Origin of Species by Means of Natural Selection*, Chapter 2; Wheeler, "Taxonomic Triage and the Poverty of Phylogeny."

63 Simon Fraser University, Department of Archaeology, "The Netloft at Namu."

64 Theresa May (caretaker at Namu), personal communication with author, July 2012.

65 Haig-Brown, *To Know a River*, 105.

66 R.G. Matson, personal communication with author, August 2010; Matson, *The Glenrose Cannery Site*; Matson, "The Old Cordilleran Component at the Glenrose Cannery Site."

67 R.G. Matson, personal communication with author, August 2010; Matson, *The Glenrose Cannery Site*; Matson, "The Old Cordilleran Component at the Glenrose Cannery Site"; Matson, "Prehistoric Subsistence Patterns in the Fraser Delta"; Carlson, "The Early Component at Bear Cove"; Prentiss, "The Emergence of New Socioeconomic Strategies."

68 Cannon and Yang, "Early Storage and Sedentism on the Pacific Northwest Coast"; Thornton et al., "Local and Traditional Knowledge"; Moss and Cannon, *The Archaeology of North Pacific Fisheries.*

69 Mackie, "Glenrose Cannery Under Threat?"; Hutchings and Campbell, "The Importance of Deltaic Wetland Resources."

70 Brian Thom (anthropologist, University of Victoria), personal communication with author, June 6, 2011.

71 Cook, *The Voyages of Captain James Cook*, 156; Johannes, "Traditional Marine Conservation Methods in Oceania and Their Demise."

72 Alifereti Tawake, personal communication with author, May 2010; Jupiter and Egli, "Ecosystem-Based Management in Fiji."

73 Stannard, *Before the Horror.*

74 John Smol, personal communication with author, 2009; Gregory-Eaves et al., "Diatoms and Sockeye Salmon (*Oncorhynchus nerka*) Population Dynamics."

75 Brian Thom, personal communication with author, June 6, 2011.

76 Atran and Medin, *The Native Mind and the Cultural Construction of Nature*, 1–11; Atran, "Taxonomic Ranks, Generic Species, and Core Memes."

77 Atran, Medin, and Ross, "The Cultural Mind."

78 Ibid.

79 Schulte, Burke, and Lipcius, "Unprecedented Restoration of a Native Oyster Metapopulation."

80 See Erlandson et al., "Paleoindian Seafaring, Maritime Technologies, and Coastal Foraging on California's Channel Islands"; Erlandson and Rick, "Archaeology Meets Marine Ecology"; McKechnie et al., "Archaeological Data Provide Alternative Hypotheses"; McKechnie, "An Archaeology of Food and Settlement on the Northwest Coast"; Lepofsky and Caldwell, "Indigenous Marine Resource Management."

81 Barnett, "The Coast Salish of Canada," 122.

82 Ibid., 124.

83 Barnett, "The Nature and Function of the Potlatch," 101, 122.

84 Trosper, *Resilience, Reciprocity and Ecological Economics*, 208.

85 Darwin, *On the Origin of Species*, Kindle, location 5783.

86 Megginson, "Lessons from Europe for American Business."

BIBLIOGRAPHY

Agriculture and Agri-food Canada. *Canadian Salmon: The Emperor of Fish*. 2011. http://publications.gc.ca/site/eng/440267/publication.html.

American Association for the Advancement of Science. "Bright Beaks Provide Honest Look at Male Immune System, *Science* Study Says." April 3, 2003. http://www.eurekalert.org/pub_releases/2003-04/aaft-bbp032803.php.

Ames, J., and S. Schroder. "Chum Salmon Colors." Washington Department of Fish and Wildlife. Last modified 2014. http://wdfw.wa.gov/fishing/salmon/chum/chum_colors.html.

Araki, H., B.A. Berejikian, M.J. Ford, and M.S. Blouin. "Fitness of Hatchery-Reared Salmonids in the Wild." *Evolutionary Applications* 1, no. 2 (2008): 342–355.

Asche, Frank, Hävard Hansen, Ragnar Tveteras, and Sigbjørn Tveterås. "The Salmon Disease Crisis in Chile." *Marine Resource Economics* 24, no. 4 (2009): 405–411.

Atran, S. "Anthropogenic Vegetation: A Garden Experiment in the Maya Lowlands." In *The Lowland Maya Area: Three Millennia at the Human-Wildland Interface*, edited by A. Gomez Pompa, M.F. Allen, S.L. Fedick, and J.J. Jiménez-Osornio, 517–532. Boca Raton, FL: CRC Press, 2003.

Atran, S. "Taxonomic Ranks, Generic Species, and Core

Memes." *Behavioral and Brain Sciences* 21, no. 4 (1998): 593–604.

Atran, S., and D. Medin. *The Native Mind and the Cultural Construction of Nature*. Boston, MA: MIT Press, 2008.

Atran, S., D. Medin, and N. Ross. "The Cultural Mind: Environmental Decision Making and Cultural Modeling Within and Across Populations." *Psychological Review* 112, no. 4 (2005): 744–776.

Atran, S., D.L. Medin, and N.O. Ross. "Evolution and Devolution of Knowledge: A Tale of Two Biologies." *Journal of the Royal Anthropological Institute* 10, no. 2 (2004): 395–420.

Avery, G. "Discussion on the Age and Use of Tidal Fish-Traps (visvywers)." *The South African Archaeological Bulletin* 30, (1975): 105–113.

Barnett, H.G. "The Coast Salish of Canada." *American Anthropologist* 40, no. 1 (1938): 118–141.

Barnett, Homer. "The Nature and Function of the Potlatch." PhD diss., University of Oregon, 1938.

BC Pacific Salmon Forum. *Final Report and Recommendations to the Government of British Columbia, January 2009*. http://www.marineharvestcanada.com/documents/BCPSFFinRptqSm.pdf

Beacham, T.D., B. McIntosh, C. MacConnachie, K.M. Miller, R.E. Withler, and N.V. Varnavskaya. "Pacific Rim Population Structure of Sockeye Salmon as Determined from Microsatellite Analysis." *Transactions of the American Fisheries Society* 135, no. 1 (2006): 174–187.

Beacham, T.D., J.R. Candy, K.D. Le, and M. Wetklo. "Population Structure of Chum Salmon (*Oncorhynchus keta*) across the Pacific Rim, Determined from Microsatellite Analysis." *Fishery Bulletin* 107, no. 2 (2009): 244–260.

Beacham, T.D., K.L. Jonsen, J. Supernault, M. Wetklo, L. Deng, and N. Varnavskaya. "Pacific Rim Population Structure of Chinook Salmon as Determined from Microsatellite Analysis." *Transactions of the American Fisheries Society* 135, no. 6 (2006): 1604–1621.

Beall, Cynthia M. "Andean, Tibetan, and Ethiopian Patterns of Adaptation to High-Altitude Hypoxia." *Integrative and Comparative Biology* 46, no. 1 (2006): 18–24.

Beall, Cynthia M., Michael Decker, Gary M. Brittenham, Irving Kushner, Amha Gebremedhin, Kingman P. Strohl. "An Ethiopian Pattern of Human Adaptation to High-Altitude Hypoxia." *Proceedings of the National Academy of Sciences of the United States of America* 99, no. 16 (2002): 17215–17218.

van den Berghe, L.P., and M.R. Gross. "Female Size and Nest Depth in Coho Salmon (*Oncorhynchus kisutch*)." *Canadian Journal of Fisheries and Aquatic Sciences* 41, no. 1 (1984): 204–206.

Blount, Jonathan D., Neil B. Metcalfe, Kathryn E. Arnold, Peter F. Surai, Godefroy L. Devevey, and Pat Monaghan. "Neonatal Nutrition, Adult Antioxidant Defences and Sexual Attractiveness in Zebra Finch." *Proceedings of the Royal Society B: Biological Sciences* 270 (2003): 1691–1696.

Boas, Franz. *Indian Myths & Legends from the North Pacific Coast of America*. Edited by Randy Bouchard and Dorothy Kennedy. Translated by Dietrich Bertz. Vancouver, BC: Talonbooks, 2002.

Braun, D.R., J.W. Harris, N.E. Levin, J.T. McCoy, A.I. Herries, M.K. Bamford, L.C. Bishop, B.G. Richmond, and M. Kibunjia. "Early Hominin Diet Included Diverse Terrestrial and Aquatic Animals 1.95 Ma in East Turkana, Kenya." *Proceedings of the National Academy of Sciences* 107, no. 22 (2010): 10002–10007.

British Library. "Treasures in Full Magna Carta." Accessed May 15, 2014. http://www.bl.uk/treasures/magnacarta/translation/mc_trans.html.

Butler, V.L., and S.K. Campbell. "Resource Intensification and Resource Depression in the Pacific Northwest of North America: A Zooarchaeological Review." *Journal of World Prehistory* 18, no. 4 (2004): 327–405.

Caldwell, M. "Fish Traps and Shell Middens at Comox Harbour, British Columbia." In *The Archaeology of North Pacific Fisheries*, edited by Madonna L. Moss and Aubrey Cannon, 235–245. Fairbanks: University of Alaska Press, 2011.

Caldwell, M. "A View from the Shore: Interpreting Fish Trap Use in Comox Harbour through Zooarchaeological Analysis of Fish Remains from the Q'umu?xs Village Site (DkSf-19), Comox Harbour, British Columbia." MA thesis, University of Manitoba, 2008.

Caldwell, M.E., D. Lepofsky, G. Combes, M. Washington, J.R. Welch, and J.R. Harper. "A Bird's Eye View of Northern

Coast Salish Intertidal Resource Management Features, Southern British Columbia, Canada." *The Journal of Island and Coastal Archaeology* 7, no. 2 (2012): 219–233.

Campbell, S.K., and V.L. Butler. "Archaeological Evidence for Resilience of Pacific Northwest Salmon Populations and the Socioecological System over the Last ~7,500 Years." *Ecology and Society* 15, no. 1 (2010): 17. http://www.ecologyandsociety.org/vol15/iss1/art17/.

Cannon, Aubrey, and Dongya Y. Yang. "Early Storage and Sedentism on the Pacific Northwest Coast: Ancient DNA Analysis of Salmon Remains from Namu, British Columbia." *American Antiquity* 71, no. 1 (2006): 123–140.

Carlson, Catherine. "The Early Component at Bear Cove." *Canadian Journal of Archaeology/Journal Canadien d'Archéologie* (1979): 177–194.

Census of Marine Life. "Niskin Bottles." Accessed May 15, 2014. http://www.coml.org/edu/tech/collect/niskin.htm

Christie, K.S., and T.E. Reimchen. "Post-Reproductive Pacific Salmon, *Oncorhynchus* spp., as a Major Nutrient Source for Large Aggregations of Gulls, *Larus* spp." *The Canadian Field-Naturalist* 119, no. 2 (2005): 202–207.

Clews, E., I. Durance, I.P. Vaughan, and S.J. Ormerod. "Juvenile Salmonid Populations in a Temperate River System Track Synoptic Trends in Climate." *Global Change Biology* 16, no. 12 (2010): 3271–3283.

Connaway, John M. *Fishweirs: A World Perspective with Emphasis on the Fishweirs of Mississippi*. Biloxi, MS: Mississippi Department of Archives and History, 2007.

Cook, James. *The Voyages of Captain James Cook*. Volume 2. London: William Smith, 1842.

Cooke, S.J., S.G. Hinch, G.T. Crossin, D. Patterson, K.K. English, J.M. Shrimpton, G. Van Der Krak, and A.P. Farrell. "Physiology of Individual Late-Run Fraser River Sockeye Salmon (*Oncorhynchus nerka*) Sampled in the Ocean Correlates with Fate during Spawning Migration." *Canadian Journal of Fisheries and Aquatic Sciences* 63 (2006): 1469–1480.

Cooney, Gabriel. "Introduction: Seeing Land from the Sea." *World Archaeology* 35, no. 3 (2004): 323–328.

Crozier, L.G., R.W. Zabel, and A.F. Hamlet. "Predicting Differential Effects of Climate Change at the Population Level with Life-Cycle Models of Spring Chinook Salmon." *Global Change Biology* 14, no. 2 (2008): 236–249.

Cullon, D., and H. Pratt. *Laich-Kwil-Tach Fish Trap Study 2010: Final Report*. Heritage Inspection Permit# 2010–0150. 2011.

Curtis, Edward S. *The North American Indian: The Complete Portfolios*. New York: Taschen, 1997.

Danziger, Danny, and John Gillingham. *1215: The Year of the Magna Carta*. London: Hodder & Stoughton, 2003.

Darwin, Charles. *The Life and Letters of Charles Darwin*. Volume 1. Edited by Francis Darwin. Project Gutenberg. Accessed May 15, 2014. http://www.gutenberg.org/files/2087/2087-h/2087-h.htm.

Darwin, Charles. *On the Origin of Species, A Facsimile of the First Edition*. Cambridge, MA: Harvard University Press, 2009. First published 1859.

Darwin, Charles. *On the Origin of Species by Means of Natural Selection*. Project Gutenberg (Kindle Edition). http://www.gutenberg.org/ebooks/1228?msg=welcome_stranger.

Dawson, T. *Locating Fish Traps on the Moray and the Forth*. 2004. Scottish Coastal Archaeology and the Problem of Erosion (SCAPE). http://www.scapetrust.org/pdf/Fish%20traps/fishtraps1.pdf

DeLoach, D.B. "The Salmon Canning Industry." Oregon State Monographs. Economic Studies Number 1, February 1939. Accessed June 2, 2014. http://ir.library.oregonstate.edu/xmlui/bitstream/handle/1957/35333/TheSalmonCanningIndustry.pdf?sequence=1.

Dietz, Thomas, Elinor Ostrom, and Paul C. Stern "The Struggle to Govern the Commons." *Science* 302, no. 5652 (December 2003): 1907–1912.

Duff, W., and British Columbia Provincial Museum. *The Indian History of British Columbia*. Victoria: Provincial Museum of Natural History and Anthropology, 1969.

Duff, W., and British Columbia Provincial Museum. *The Upper Stalo Indians of the Fraser Valley*. Victoria: Provincial Museum of Natural History and Anthropology, 1953.

Eliason, Erika J., Timothy D. Clark, Merran J. Hague, Linda M. Hanson, Zoë S. Gallagher, Ken M. Jeffries, Marika K. Gale, David A. Patterson, Scott G. Hinch, and Anthony P. Farrell. "Differences in Thermal Tolerance among Sockeye Salmon Populations." *Science* 332, no. 6025 (April 2011): 109–112. doi:10.1126/science.1199158.

Encyclopædia Britannica. "Niskin Bottle." Accessed May 15, 2014. http://www.britannica.com/EBchecked/topic/1660258/Niskin-bottle.

Encyclopedia of Life. "*Oncorhynchus gorbuscha*." Accessed April 11, 2014. http://eol.org/pages/205246/overview.

English Heritage. *Introductions to Heritage Assets: River Fisheries and Coastal Fish Weirs*. May 2011. Accessed April 11, 2014. https://www.english-heritage.org.uk/publications/iha-river-fisheries-coastal-fish-weirs/riverfisheriescoastalfishweirs.pdf.

Ereshefsky, Marc. "Darwin's Solution to the Species Problem." *Synthese* 175, no. 3 (2010): 405–425.

Erickson, Clark L. "The Domesticated Landscapes of the Bolivian Amazon." In *Time and Complexity in Historical Ecology: Studies in the Neotropical Lowlands*, edited by William L. Balée and Clark L. Erickson, 235–278. New York: Columbia University Press, 2006.

Erlandson, J.M., and T.C. Rick. "Archaeology Meets Marine Ecology: The Antiquity of Maritime Cultures and Human Impacts on Marine Fisheries and Ecosystems." *Annual Review of Marine Science* 2 (2010): 165–185.

Erlandson, J.M., T.C. Rick, T.J. Braje, M. Casperson, B. Culleton, B. Fulfrost, T. Garcia, D.A. Guthrie, N. Jew, D.J. Kennett, M.L. Moss, L. Reeder, C. Skinner, J. Watts, and L. Willis. "Paleoindian Seafaring, Maritime Technologies, and Coastal Foraging on California's Channel Islands." *Science* 331, no. 6021 (2011): 1181–1185.

Farrell, A.P., S.G. Hinch, S.J. Cooke, D.A. Patterson, G.T.

Crossin, M. Lapointe, and M.T. Mathes. "Pacific Salmon in Hot Water: Applying Aerobic Scope Models and Biotelemetry to Predict the Success of Spawning Migrations." *Physiological and Biochemical Zoology* 81, no. 6 (2008): 697–709.

Ferrari, Michael R., James R. Miller, and Gary L. Russell. "Modeling Changes in Summer Temperature of the Fraser River during the Next Century." *Journal of Hydrology* 342, no. 3–4 (2007): 336–346.

Field, R.D., and J.D. Reynolds. "Sea to Sky: Impacts of Residual Salmon-Derived Nutrients on Estuarine Breeding Bird Communities." *Proceedings of the Royal Society B: Biological Sciences* 278, no. 1721 (2011): 3081–3088.

Finney, B.P., I. Gregory Eaves, J. Sweetman, M.S. Douglas, and J.P. Smol. "Impacts of Climatic Change and Fishing on Pacific Salmon Abundance over the Past 300 Years." *Science* 290, no. 5492 (2000): 795 – 799.

Finney, B.P., I. Gregory Eaves, M.S. Douglas, and J.P. Smol. "Fisheries Productivity in the Northeastern Pacific Ocean over the Past 2,200 Years." *Nature* 416, no. 6882 (2002): 729–733.

Fisheries and Oceans Canada. *Canada's Policy for Conservation of Wild Pacific Salmon*. June 2005. http://www.pac.dfo-mpo.gc.ca/publications/pdfs/wsp-eng.pdf.

Ford, J.S., and R.A. Myers. "A Global Assessment of Salmon Aquaculture Impacts on Wild Salmonids." *PLoS biology* 6, no. 2 (2008): e33.

Fraser, D.J. "How Well Can Captive Breeding Programs Conserve Biodiversity? A Review of Salmonids." *Evolutionary Applications* 1, no. 4 (2008): 535–586.

Gende, Scott M., Richard T. Edwards, Mary F. Willson, and Mark S. Wipfli. "Pacific Salmon in Aquatic and Terrestrial Ecosystems." *BioScience* 52, no. 10 (October 2002): 917–928.

Government of British Columbia, Ministry of Agriculture, Policy and Industry Competitiveness Branch. *2010 Year in Review: British Columbia Seafood Industry*. October 2011. http://www.agf.gov.bc.ca/stats/YinReview/Seafood-YIR-2010.pdf

Grayson, Donald K., and David J. Meltzer. "A Requiem for North American Overkill." *Journal of Archaeological Science* 30, no. 5 (2003): 585–593.

Gregory-Eaves, I., D.T. Selbie, J.N. Sweetman, B.P. Finney, and J.P. Smol. "Tracking Sockeye Salmon Population Dynamics from Lake Sediment Cores: A Review and Synthesis." *American Fisheries Society Symposium* 69 (August 2009): 379–393.

Gregory-Eaves, I., J.P. Smol, M.S. Douglas, and B.P. Finney. "Diatoms and Sockeye Salmon (*Oncorhynchus nerka*) Population Dynamics: Reconstructions of Salmon-Derived Nutrients over the Past 2,200 Years in Two Lakes from Kodiak Island, Alaska." *Journal of Paleolimnology* 30, no. 1 (2003): 35–53.

Groot, Cornelis, and Leo Margolis, eds. *Pacific Salmon Life Histories*. Vancouver: UBC Press, 1991.

Haig-Brown, Roderick L. *Fisherman's Fall*. Toronto: Collins, 1964.

Haig-Brown, Roderick L. *Fisherman's Spring*. Toronto: Collins, 1951.

Haig-Brown, Roderick L. *Fisherman's Summer*. New York: W. Morrow, 1959.

Haig-Brown, Roderick L. *Fisherman's Winter*. New York: W. Morrow, 1954.

Haig-Brown, Roderick L. *Return to the River, A Story of the Chinook Run*. Toronto: Collins, 1941.

Haig-Brown, Roderick L. *A River Never Sleeps*. New York: W. Morrow, 1946.

Haig-Brown, Roderick L. *To Know a River*. Guilford, CT: Globe Pequot Press, 2000

Hayashida, Frances M. "Archaeology, Ecological History, and Conservation." *Annual Review of Anthropology* 34 (2005): 43–65.

Hendry A.P., J.K. Wenburg, P. Bentzen, E.C. Volk, and T.P. Quinn. "Rapid Evolution of Reproductive Isolation in the Wild: Evidence from Introduced Salmon." *Science* 290, no. 5491 (October 20, 2000): 516–519.

Hinch, S.G., and E.G. Martins. *A Review of Potential Climate Change Effects on Survival of Fraser River Sockeye Salmon and an Analysis of Interannual Trends in En Route Loss and Pre-Spawn Mortality*. Cohen Commission Technical Report 9. Vancouver, BC: Cohen Commission, 2011. http://www.watershed-watch.org/wordpress/wp-content/uploads/2011/06/Exh-553-NonRT.pdf.

Hinch, S.G., and J. Gardner, eds. *Conference on Early Migration and Premature Mortality in Fraser River Late-Run Sockeye Salmon: Proceedings Forest Sciences Centre, UBC, June 16 to 18, 2008*. Vancouver: Pacific Fisheries Resource Conservation Council, 2009.

Hine, P., J. Sealy, D. Halkett, and T. Hart. "Antiquity of Stone-Walled Tidal Fish Traps on the Cape Coast, South Africa." *The South African Archaeological Bulletin* (2010): 35–44.

Hocking, M.D., and J.D. Reynolds. "Impacts of Salmon on Riparian Plant Diversity." *Science* 331, no. 6024 (2011): 1609–1612.

Hocking, M.D., and J.D. Reynolds. "Nitrogen Uptake by Plants Subsidized by Pacific Salmon Carcasses: A Hierarchical Experiment." *Canadian Journal of Forest Research* 42, no. 5 (2012): 908–917.

Hocking, M.D., R.A. Ring, and T.E. Reimchen. "Burying Beetle *Nicrophorus investigator* Reproduction on Pacific Salmon Carcasses." *Ecological Entomology* 31, no. 1 (2006): 5–12.

Hoffmann, Richard C. "A Brief History of Aquatic Resource Use in Medieval Europe." *Helgoland Marine Research* 59, no. 1 (2005): 22–30.

Hoffmann, Richard C. "Economic Development and Aquatic Ecosystems in Medieval Europe." *The American Historical Review* 101, no. 3 (June 1996): 631–669.

Hunter, J.G. "Survival and Production of Pink and Chum Salmon in a Coastal Stream." *Journal of the Fisheries Board of Canada* 16, no. 6 (1959): 835–886.

Hutchings, Jeffrey A. "Spatial and Temporal Variation in the Density of Northern Cod and a Review of Hypotheses for the Stock's Collapse." *Canadian Journal of Fisheries and Aquatic Sciences* 53, no. 5 (1996): 943–962.

Hutchings, Richard M., and Sarah K. Campbell. "The Importance of Deltaic Wetland Resources: A Perspective from the Nooksack River Delta, Washington State, USA." *Journal of Wetland Archaeology* 5, no. 1 (2005): 17–34.

Huxley, Camilla. "Symbiosis between Ants and Epiphytes." *Biological Reviews* 55, no. 3 (1980): 321–340.

Irvine, J.R., R.W. Macdonald, R.J. Brown, L. Godbout, J.D. Reist, and E.C. Carmack. "Salmon in the Arctic and How They Avoid Lethal Low Temperatures." *North Pacific Anadromous Fish Commission Bulletin* 5 (2009): 39–50.

Isabella, Jude. "Wrens May Have Salmon to Thank for Breeding Boost." *New Scientist*, October 4, 2012. Accessed June 5, 2014. http://www.newscientist.com/article/dn22341-wrens-may-have-salmon-to-thank-for-breeding-boost.html#.U31GqCggUQ5.

Johannes, Robert E. "Traditional Marine Conservation Methods in Oceania and Their Demise." *Annual Review of Ecology and Systematics* 9, no. 1 (1978): 349–364.

Jupiter, Stacy D., and Daniel P. Egli. "Ecosystem-Based Management in Fiji: Successes and Challenges after Five Years of Implementation." *Journal of Marine Biology* 2011 (2011). http://dx.doi.org/10.1155/2011/940765.

Karpenko, V.I., and M.V. Koval. "Feeding Strategies and Trends of Pink and Chum Salmon Growth in the Marine

Waters of Kamchatka." *North Pacific Anadromous Fish Commission Technical Report* 8 (2012): 82–86.

Kitada, S., H. Shishidou, T. Sugaya, T. Kitakado, K. Hamasaki, and H. Kishino. "Genetic Effects of Long-Term Stock Enhancement Programs." *Aquaculture* 290, no. 1 (2009): 69–79.

Knapp, G. "Trends in Pink and Chum Salmon Markets: What the Data Show." Paper presented at the Pink and Chum Salmon Workshop, Juneau, Alaska, February 14, 2012. http://pinkandchum.psc.org/Presentation/Knapp.pdf.

Knapp, G. "The Wild Salmon Industry: Five Predictions for the Future." *Fisheries Economics Newsletter* 51 (May 2001). Accessed May 15, 2014. http://www.iser.uaa.alaska.edu/people/knapp/personal/Wild%20salmon%20future.pdf.

Kocan, R., and P. Hershberger. "Differences in *Ichthyophonus* Prevalence and Infection Severity between Upper Yukon River and Tanana River Chinook Salmon, *Oncorhynchus tshawytscha* (Walbaum), stocks." *Journal of Fish Diseases* 29, no. 8 (2006): 497–503.

Kocan, Richard, Paul Hershberger, and James Winton. "Ichthyophoniasis: An Emerging Disease of Chinook Salmon in the Yukon River." *Journal of Aquatic Animal Health* 16, no. 2 (2004): 58–72.

Kovach, Ryan P., Anthony J. Gharrett, and David A. Tallmon. "Genetic Change for Earlier Migration Timing in a Pink Salmon Population." *Proceedings of the Royal Society B: Biological Science*. July 2012. http://rspb.

royalsocietypublishing.org/content/early/2012/07/03/rspb.2012.1158.full.

Krkošek, M., J.S. Ford, A. Morton, S. Lele, R.A. Myers, and M.A. Lewis. "Declining Wild Salmon Populations in Relation to Parasites from Farm Salmon." *Science* 318, no. 5857 (2007): 1772–1775.

Lackey, Robert T. "Salmon Decline in Western North America: Historical Context." In *Encyclopedia of Earth*, edited by Cutler J. Cleveland. Washington, DC: Environmental Information Coalition, National Council for Science and the Environment, 2008. http://www.epa.gov/wed/pages/staff/lackey/pubs/ENCYCLOPEDIA-OF-EARTH-LACKEY-SALMON-HISTORY-MS-2008.pdf.

Lathan, Robert S. "Rubber Gloves Redux." *Proceedings (Baylor University Medical Center)* 24, no. 4 (October 2011): 324. http://www.ncbi.nlm.nih.gov/pmc/articles/PMC3205159/.

Lepofsky, D., K. Lertzman, D. Hallett, and R. Mathewes. "Climate Change and Culture Change on the Southern Coast of British Columbia 2400–1200 Cal. BP: An Hypothesis." *American Antiquity* 70, no. 2 (2005): 267–293.

Lepofsky, D., and M. Caldwell. "Indigenous Marine Resource Management on the Northwest Coast of North America." *Ecological Processes* 2, no. 1 (2013): 1–12.

Lewes, John Lee. "Crisis in the Fur Trade." In *Trading Beyond the Mountains: The British Fur Trade on the Pacific, 1793–1843*, by Richard S. Mackie, 244. Vancouver: UBC Press, 1997.

Mackie, Quentin. "Glenrose Cannery Under Threat?" *Northwest Coast Archaeology*. October 2, 2010. Accessed May 15, 2014. http://qmackie.com/2010/10/02/glenrose-cannery-under-threat/.

Mackie, Quentin, Daryl Fedje, Duncan McLaren, Nicole Smith, and Iain McKechnie. "Early Environments and Archaeology of Coastal British Columbia." In *Trekking the Shore*, edited by Nuno F. Bicho, J.A. Haws, and L.G. Davis, 51–103. New York: Springer, 2011.

Mackinlay, Don, Susan Lehmann, Joan Bateman, and Roberta Cook. *Pacific Salmon Hatcheries in British Columbia*. Vancouver: Fisheries and Oceans Canada. Accessed May 15, 2014. http://www.sehab.ca/pdf/hatcheries.pdf.

Maclachlan, M., and W.P. Suttles. *The Fort Langley Journals, 1827–30*. Vancouver: UBC Press, 1998.

"The Magna Carta Project." Accessed May 15, 2014. http://magnacarta.cmp.uea.ac.uk/read/magna_carta_1215/!all?commentary=secondary.

Mantua, N.J. "Patterns of Change in Climate and Pacific Salmon Production." *American Fisheries Society Symposium* 70 (2009): 1143–1157.

Marine Stewardship Council. "Three British Columbia Chum Salmon Fisheries Awarded MSC Certification." January 8, 2013. Accessed May 15, 2014. http://www.msc.org/newsroom/news/three-british-columbia-chum-salmon-fisheries-awarded-msc-certification.

Martin, Paul. "The Discovery of America." *Science* 179, no. 4077 (1973): 969–974.

Martin, Paul S. "Prehistoric Overkill: The Global Model." In *Quaternary Extinctions: A Prehistoric Revolution*, edited by Paul S. Martin and Richard G. Klein, 354–403. Tucson: University of Arizona Press, 1984.

Martins, Eduardo G., Scott G. Hinch, David A. Patterson, Merran J. Hague, Steven J. Cooke, Kristina M. Miller, Michael F. Lapointe, Karl K. English, and Anthony P. Farrell. "Effects of River Temperature and Climate Warming on Stock-Specific Survival of Adult Migrating Fraser River Sockeye Salmon (*Oncorhynchus nerka*)." *Global Change Biology* 17, no. 1 (2011): 99–114.

Matson, R.G. *The Glenrose Cannery Site*. Ottawa: National Museums of Canada, 1976.

Matson, R.G. "The Old Cordilleran Component at the Glenrose Cannery Site." In *Early Human Occupation in British Columbia*, edited by Roy Carlson and Luke Dalla Bona, 111–122. Vancouver: UBC Press, 1996..

Matson, R.G. "Prehistoric Subsistence Patterns in the Fraser Delta: The Evidence from the Glenrose Cannery Site." *BC Studies: The British Columbian Quarterly* 48 (1980): 64–85.

McGraw, Kevin J., and Daniel R. Ardia. "Carotenoids, Immunocompetence, and the Information Content of Sexual Colors: An Experimental Test." *The American Naturalist* 162, no. 6 (December 2003): 704–712. http://www.jstor.org/stable/10.1086/378904.

McKechnie, Iain. "An Archaeology of Food and Settlement on the Northwest Coast." PhD diss., University of British Columbia, 2013.

McKechnie, Iain, Dana Lepofsky, Madonna L. Moss, Virginia L. Butler, Trevor J. Orchard, Gary Coupland, Fredrick Foster, Megan Caldwell, and Ken Lertzman. "Archaeological Data Provide Alternative Hypotheses on Pacific Herring (*Clupea pallasii*) Distribution, Abundance, and Variability." *Proceedings of the National Academy of Sciences of the United States of America* 111, no. 9 (2014): E807–E816.

Megginson, Leon C. "Lessons from Europe for American Business." *Southwestern Social Science Quarterly* 44, no. 1 (1963): 3–13.

Meggs, G. *Salmon: The Decline of the British Columbia Fishery*. Vancouver: Douglas & McIntyre, 1991.

Miller, K.A. "Pacific Salmon Fisheries: Climate, Information and Adaptation in a Conflict-Ridden Context." In *Societal Adaptation to Climate Variability and Change*, edited by S. Kane and G. Yohe, 37–61. The Netherlands: Springer, 2000.

Miller, K.M., S. Li, K.H. Kaukinen, N. Ginther, E. Hammill, J.M.R. Curtis, D.A. Patterson, T. Sierocinski, L. Donnison, P. Pavlidis, S.G. Hinch, K.A. Hruska, S.J. Cooke, K.K. English, and A.P. Farrell. "Genomic Signatures Predict Migration and Spawning Failure in Wild Canadian Salmon." *Science* 331, no. 6014 (2011): 214–217.

Miyakoshi, Yasuyuki, Mitsuhiro Nagata, Shuichi Kitada, and

Masahide Kaeriyama. "Historical and Current Hatchery Programs and Management of Chum Salmon in Hokkaido, Northern Japan." *Reviews in Fisheries Science* 21, no. 3–4 (2003): 469–479.

Morita, K., S.H. Morita, and M.A. Fukuwaka. "Population Dynamics of Japanese Pink Salmon (*Oncorhynchus gorbuscha*): Are Recent Increases Explained by Hatchery Programs or Climatic Variations?" *Canadian Journal of Fisheries and Aquatic Sciences* 63, no. 1 (2006): 55–62.

Moss, Madonna L., and Aubrey Cannon, eds. *The Archaeology of North Pacific Fisheries.* Fairbanks: University of Alaska Press, 2011.

Nadasdy, P. *Hunters and Bureaucrats: Power, Knowledge, and Aboriginal-State Relations in the Southwest Yukon.* Vancouver: UBC Press, 2003.

NASA. "Spinoff Database Record. Spinoff from a Moon Tool." Last modified May 1, 2011. Accessed May 15, 2014. http://spinoff.nasa.gov/spinoff/spinitem?title=Spinoff+from+a+Moon+Tool.

National Archives. "Salmon and Freshwater Fisheries Act 1975." Accessed May 15, 2014. http://www.legislation.gov.uk/ukpga/1975/51.

Noakes, D.J., R.J. Beamish, and R. Gregory. "British Columbia's Commercial Salmon Industry." NPAFC Doc. No. 642. Nanaimo, BC: Department of Fisheries and Oceans, Sciences Branch, Pacific Region, Pacific Biological Station, 2002.

North Pacific Anadromous Fish Commission. "Rehabilitation of Japanese Salmon Hatcheries Hit by the March 11, 2011, Earthquake and Tsunami." NPAFC Newsletter No. 32, July 2012. Accessed May 15, 2014. http://www.npafc.org/new/publications/Newsletter/NL32/Newsletter32.pdf.

Northwest Fisheries Science Center. "Copepod Biodiversity." Accessed May 15, 2014. http://www.nwfsc.noaa.gov/research/divisions/fe/estuarine/oeip/ea-copepod-biodiversity.cfm.

Omega Pacific. "Culturing Pacific Salmon since 1980." Accessed May 15, 2014. http://www.omegapacific.ca/modules/tinyd1/index.php?id=1/.

O'Sullivan, A. "Exploring Past People's Interactions with Wetland Environments in Ireland." *Proceedings of the Royal Irish Academy, Section C* 107, no. 1 (January 2007): 147–203.

Pacific Salmon Commission. *Report of the Fraser River Panel to the Pacific Salmon Commission on the 2009 Fraser River Sockeye and Pink Salmon Fishing Season*. March 2013. ftp://ftp.psc.org/pub/SEF_pink_salmon/frpar2009_v2.pdf.

Parenti, P. "On the Status of Species Classified in the Genus *Perca* by Johann Julius Walbaum (1792)." *Zoological Studies (Taipei)* 42, no. 4 (2003): 491–505.

Parks Canada. "Fort Langley National Historic Site of Canada." Accessed May 15, 2014. http://www.pc.gc.ca/eng/lhn-nhs/bc/langley/natcul/natcul2/a.aspx#econom.

Pierson, N. "Bridging Troubled Waters: Zooarchaeology and

Marine Conservation on Burrard Inlet, Southwest British Columbia." PhD diss., Simon Fraser University, 2011.

Prentiss, Anna Marie. "The Emergence of New Socioeconomic Strategies in the Middle and Late Holocene Pacific Northwest Region of North America." In *Macroevolution in Human Prehistory: Evolutionary Theory and Processual Archaeology*, edited by Anna Prentiss, Ian Kuijt, and James C. Chatters, 111–131. New York: Springer, 2009.

Pringle, H. "Herring and Nuts for the 'Salmon People.'" *Science* 320, no. 5873 (2008): 174–175.

Quinn, Thomas P. *The Behavior and Ecology of Pacific Salmon and Trout*. Vancouver: UBC Press, 2005.

Quinn, Thomas P., Ronald T. Merrill, and Ernest L. Brannon. "Magnetic Field Detection in Sockeye Salmon." *Journal of Experimental Zoology* 217, no. 1 (1981): 137–142.

Raby, G.D., S.J. Cooke, K.V. Cook, S.H. McConnachie, M.R. Donaldson, S.G. Hinch, C.K. Whitney, S.M. Drenner, D.A. Patterson, T.D. Clark, and A.P. Farrell. "Resilience of Pink Salmon and Chum Salmon to Simulated Fisheries Capture Stress Incurred upon Arrival at Spawning Grounds." *Transactions of the American Fisheries Society* 142, no. 2 (2013): 524–539.

Reimchen, T. "Some Considerations in Salmon Management." In *Ghost Runs: The Future of Wild Salmon on the North and Central Coasts of British Columbia*, edited by B. Harvey and M. MacDuffee, 93–96. Victoria, BC: Raincoast Conservation Foundation, 2002.

Riddell, B., and Pacific Fisheries Resource Conservation Council. *Pacific Salmon Resources in Central and North Coast British Columbia*. Vancouver, BC: Pacific Fisheries Resource Conservation Council, 2004.

"Ripple Rock Explosion 1958." Youtube. Uploaded August 16, 2009. Accessed May 15, 2014. https://www.youtube.com/watch?v=xOh20xpFmZI.

Roberts, Callum. *The Unnatural History of the Sea*. Washington, DC: Island Press, 2007.

Rose, G.A. "Reconciling Overfishing and Climate Change with Stock Dynamics of Atlantic Cod (*Gadus morhua*) over 500 Years." *Canadian Journal of Fisheries and Aquatic Sciences* 61, no. 9 (2004): 1553–1557.

Ruggerone, G.T., M. Zimmermann, K.W. Myers, J.L. Nielsen, and D.E. Rogers. "Competition between Asian Pink Salmon (*Oncorhynchus gorbuscha*) and Alaskan Sockeye Salmon (*O. nerka*) in the North Pacific Ocean." *Fisheries Oceanography* 12, no. 3 (2003): 209–219.

Scheinfeldt, Laura B., Sameer Soi, Simon Thompson, Alessia Ranciaro, Dawit Woldemeskel, William Beggs, Charla Lambert, Joseph P. Jarvis, Dawit Abate, Gurja Belay, and Sarah A. Tishkoff. "Genetic Adaptation to High Altitude in the Ethiopian Highlands." *Genome Biology* 13, no. 1 (2012): R1.

Schindler, Daniel E., Ray Hilborn, Brandon Chasco, Christopher P. Boatright, Thomas P. Quinn, Lauren A. Rogers, and Michael S. Webster. "Population Diversity and the Portfolio Effect in an Exploited Species." *Nature* 465, no. 7298 (2010): 609–612.

Schroder, S.L. *Assessment of Production of Chum Salmon Fry from the Big Beef Creek Spawning Channel: Annual Report – Anadromous Fish Project*. 1975. https://digital.lib.washington.edu/researchworks/handle/1773/3829.

Schroder, S.L. "The Role of Sexual Selection in Determining Overall Mating Patterns and Mate Choice in Chum Salmon." PhD diss., University of Washington, 1981.

Schulte, David M., Russell P. Burke, and Romuald N. Lipcius. "Unprecedented Restoration of a Native Oyster Metapopulation." *Science* 325, no. 5944 (2009): 1124–1128.

Scripps Institution of Oceanography. "Scripps Oceanography Researchers Discover Arctic Blooms Occurring Earlier." March 2, 2011. https://scripps.ucsd.edu/news/1961

Seattle Post-Intelligencer. "Rainbow Trout, Dyed Red, Is Being Sold as Coho Salmon." August 21, 1994.

Selbie, D.T., B.A. Lewis, J.P. Smol, and B.P. Finney. "Long-Term Population Dynamics of the Endangered Snake River Sockeye Salmon: Evidence of Past Influences on Stock Decline and Impediments to Recovery." *Transactions of the American Fisheries Society* 136, no. 3 (2007): 800–821.

Simon Fraser University, Department of Archaeology. "The Netloft at Namu." http://www.sfu.ca/archaeology-old/museum/bc/namu_src/NH000002.HTM.

Spilsted, B., G. Pestal, and Department of Fisheries and Oceans. *Certification Unit Profile: North Coast and Central Coast Chum Salmon*. Vancouver: Fisheries

& Aquaculture Management Branch, Department of Fisheries and Oceans, 2009.

Stannard, David E. *Before the Horror: The Population of Hawai'i on the Eve of Western Contact*. Honolulu: Social Science Research Institute, University of Hawaii, 1989.

State of Alaska, Alaska Department of Fish and Game. "Pink Salmon (*Oncorhynchus gorbuscha*)." Accessed May 15, 2014. http://www.adfg.alaska.gov/index.cfm?adfg=pink-salmon.printerfriendly.

Steinbeck, John. *The Log from the Sea of Cortez*. New York: Penguin Classics, 1969. First published 1951 by Viking Press.

Stewart, H. *Indian Fishing: Early Methods on the Northwest Coast*. Vancouver: J.J. Douglas, 1977.

Stewart, H. *Stone, Bone, Antler & Shell: Artifacts of the Northwest Coast*. Vancouver: Douglas & McIntyre, 1996.

Thornton, Thomas F., Madonna L. Moss, Virginia L. Butler, Jamie Hebert, and Fritz Funk. "Local and Traditional Knowledge and the Historical Ecology of Pacific Herring in Alaska." *Journal of Ecological Anthropology* 14, no. 1 (2010): 81–88.

Trosper, R.L. "Resilience in Pre-contact Pacific Northwest Social Ecological Systems." *Conservation Ecology* 7, no. 3 (2003): 6. http://www.consecol.org/vol7/iss3/art6/.

Trosper, R.L. *Resilience, Reciprocity and Ecological Economics: Northwest Coast Sustainability*. London: Routledge, 2009.

Tsuda, Y., R. Kawabe, H. Tanaka, Y. Mitsunaga, T. Hiraishi, K. Yamamoto, and K. Nashimoto. "Monitoring the Spawning Behaviour of Chum Salmon with an Acceleration Data Logger." *Ecology of Freshwater Fish* 15, no. 3 (2006): 264–274.

Van Duyn Southworth, Gertrude. *A First Book in American History with European Beginnings*. New York: D. Appleton and Company, 1917. Accessed May 15, 2014. https://archive.org/stream/firstbookinamerioosouth/firstbookinamerioosouth_djvu.txt.

Waddy, S.L., and D.E. Aiken. "Multiple Fertilization and Consecutive Spawning in Large American Lobsters, *Homarus americanus*." *Canadian Journal of Fisheries and Aquatic Sciences* 43 (1986): 2291–2294.

Walters, Carl, and Jean-Jacques Maguire. "Lessons for Stock Assessment from the Northern Cod Collapse." *Reviews in Fish Biology and Fisheries* 6, no. 2 (1996): 125–137.

Waples, R.S., G.R. Pess, and T. Beechie. "Evolutionary History of Pacific Salmon in Dynamic Environments." *Evolutionary Applications* 1, no. 2 (2008): 189–206.

Washington Department of Fish and Wildlife. *Puget Sound Steelhead Technical Recovery Team Report: Identifying Historical Populations of Steelhead within the Puget Sound Distinct Population Segment*. 2013. http://wdfw.wa.gov/publications/01508/.

Welch, J., D. Lepofsky, M. Caldwell, G. Combes, and C. Rust. "Treasure Bearers: Personal Foundations for Effective Leadership in Northern Coast Salish Heritage Stewardship." *Heritage & Society* 4, no. 1 (2011): 83–114.

Wheeler, Quentin D. "Taxonomic Triage and the Poverty

of Phylogeny." *Philosophical Transactions of the Royal Society of London. Series B: Biological Sciences* 359, no. 1444 (2004): 571–583.

White/Xanius, Elroy. "Heiltsuk Stone Fish Traps: Products of My Ancestors' Labour." MA thesis, Simon Fraser University, 2006. Accessed May 15, 2014. http://summit.sfu.ca/item/4240.

Wobst, H.M. "The Archaeo-Ethnology of Hunter-Gatherers of the Tyranny of the Ethnographic Record in Archaeology." *American Antiquity (Washington, DC)* 43, no. 2 (1978): 303–309.

Wright, J.K., E.W. Carpenter, A.G. Hunt, and B. Downhill, "Observations on the Explosion at Ripple Rock." *Nature* 182, (1958): 1597–1598.

Zuray, Stanley, Richard Kocan, and Paul Hershberger. "Synchronous Cycling of Ichthyophoniasis with Chinook Salmon Density Revealed during the Annual Yukon River Spawning Migration." *Transactions of the American Fisheries Society* 141, no. 3 (2012): 615–623.

JUDE ISABELLA has been a journalist for over 20 years, focusing on science, health and the environment. She writes for a diverse audience, from grownups interested in archaeology to young readers curious about space exploration. Jude has written five science books for kids, including *Fantastic Feats and Failures*, which won the prestigious American Institute of Physics Award. She spent three years researching salmon and marine biodiversity on Canada's West Coast, resulting in this book. Jude lives in Victoria, British Columbia.